Ecosystem Functions and Management

Harpinder Sandhu
Editor

Ecosystem Functions and Management

Theory and Practice

Springer

Editor
Harpinder Sandhu
School of the Environment
Flinders University
Adelaide, SA
Australia

ISBN 978-3-319-53966-9 ISBN 978-3-319-53967-6 (eBook)
DOI 10.1007/978-3-319-53967-6

Library of Congress Control Number: 2017931346

Printed on acid-free paper

This Springer imprint is published by Springer Nature
The registered company is Springer International Publishing AG
The registered company address is: Gewerbestrasse 11, 6330 Cham, Switzerland

Foreword

The idea that biodiversity in our natural ecosystems provides important ecosystem services to humanity, and must, therefore, be protected has been gaining ground in recent decades. Perhaps nowhere else in the world, the critical role the ecosystem services play in the lives of people is more evident than in South Asia. Tens of millions of people, almost all of them from the lowest income strata- of our society, directly rely on natural ecosystems to sustain their livelihoods.

Our agricultural production depends upon several ecosystem services such as natural processes that form and maintain soils, ensure continuous flow of water for irrigation, and maintain pollinators, and other beneficial organisms that control pests and diseases. Almost everyone depends on nature to regulate air quality, and to provide aesthetic and spiritual services. Natural systems, above all, provide us options to deal with an uncertain future.

We can calculate the economic value of these ecosystem services to justify their protection and sound management, and we have done that. But we all know that nature is priceless.

Priceless, as it may be, assaults on nature continue in South Asia, as elsewhere. Some losses in ecosystem services are inevitable, as a result of development, others resulting from poor policies and governance and the lack of appreciation of full economic, social, and environmental consequences of the decline, are unnecessary. Equally important, continuing degradation ignores the inherent value of natural ecosystems in providing solutions to our social, economic, and environmental dilemmas.

This book, a contribution from some of the most well-known experts from South Asia, explores the status of ecosystem services in the region. The essays by a range of experts in ecology, sociology, and economics, describe ongoing changes and the impacts of these changes on six major ecosystems as well as societies. General trends are driven home by case studies from forest ecosystems in the mountains and agricultural systems from the plains of Punjab.

The changes in natural ecosystems are rapid, massive, and in most cases irreversible. Climate change, growing populations, and development aspirations are exacerbating these the negative impacts of these changes. Given the importance of natural ecosystems in meeting the United Nations Sustainable Development Goals, in particularly the importance of forest ecosystems in sequestering carbon, the lack of sufficient attention to the maintenance and restoration of natural systems is surprising.

Hopefully, books such as these will help draw attention to the role of natural systems in meeting society's quest for sustainable development.

Kamal Bawa
University of Massachusetts, Boston, MA, USA
Ashoka Trust for Research in Ecology and the Environment, Bangalore, India

Preface

Healthy and functioning ecosystems are necessary for the survival of human beings as they provide life-supporting goods and services. These goods and services are often called as ecosystem services (ES). ES not only provide major inputs to many sectors of the economy, but also support our lifestyles. They include biological control of pests, pollination of crops, flood control and wind erosion control, provision of food (including fisheries), carbon capture, aesthetics, etc. In other words, mankind cannot survive without healthy and functioning ecosystems. However, ecosystems and their services worldwide are being degraded more rapidly than ever before and this degradation poses serious threats to quality of life and modern economies. However, recent scientific evidence suggest that ecosystems and their services are declining rapidly ove the worldr. Asia hosts nearly one-third of the human population. To fulfill increasing needs of food and industrial products for growing population, massive industrialisation is occurring at faster pace in this region. These activities are having negative impacts on natural ecosystems and their services, which are likely to be increased in coming decades due to increasing population and their growing demands. Globally, ecosystems have been studied in much detail; however, vital information on the key local ecosystems, their functioning, state of health, etc., needs to be addressed in Asian context particularly in Indian subcontinent. This book aims to fulfill this gap by presenting current knowledge on ecosystems, their functions, values, and importance to society and the economy.

This book addresses six major ecosystems in Indian subcontinent—mountain, agroecosystems, desert, forest, urban and freshwater ecosystems. It provides an overview of their functions and discusses how they support livelihoods. Building on the ecosystems theory and literature since the United Nations Millennium Ecosystem Assessment, it then elaborates on methods, concepts, policies and practices to maintain ecosystems. It highlights the ecosystem management approach that involves management of natural resources within the ecological constraints to fulfill socio-economic objectives of the communities that are dependent on natural ecosystems. It is a holistic approach that covers social, economic and environmental aspects of the sustainability. In this book, we take anthropocentric view of the ecosystem management. It involves managing ecosystems to get desired social, cultural and economic benefits. The book concludes by discussing the significance of interlinking ecosystems for sustainable and equitable development in the region. This approach is timely and will be of high scientific and political value as discussed at the recent Rio+20 Summit to move towards green economy for sustainable and equitable development.

The book is divided into three parts. Part I is introduction and has two chapters that set the scene for rest of the book. Chapter 1 provides an overview of the different types of ecosystems in the South Asia region. It introduces an ecosystem approach including its management and value of ES. It also provides the conceptual framework of relationship between ecosystems and human well-being. This approach is further elaborated in the rest of the book chapters. Chapter 2 builds on this theme and define ecosystems and economic systems. It argues that both are evolving and interlinked. There is also need to consider ethics in designing policies to manage them for the welfare of humans and other species on earth.

Part II describes six major ecosystems in South Asia. Chapter 3 is a case study located in Himalayas as an example from mountain ecosystems. It provides an empirical exploration of interactions between humans and the environment using an ecological modelling approach. Such modelling can be used for decision-making process to manage complex ecosystems. Chapter 4 provides a case study from agroecosystems by elaborating on the agriculture in Punjab, India. It examines the current status of ES provided by agriculture in Punjab and health of water and land resources. Chapter 5 describes deserts and their management in relation to Indian subcontinent. It describes major deserts in the subcontinent with special reference to Pakistan. Chapter 6 provides overview of forest and the role they play in supporting livelihood in India. It discusses the financial incentives of improving forest ES and its impact on economy. Chapter 7 discusses urban ecosystems by taking examples from Pakistan, India and Bangladesh. It elaborates on the ES concept and its relevance to the urban ecosystem management and planning. Chapter 8 focuses on freshwater ecosystems. It provides a case study of tank water systems in India.

Part III comprises two chapters. Chapter 9 provides a framework to interlink all ecosystems discussed in the book. It also discusses the benefits of interlinkages for the long-term sustainability. Chapter 10 discusses how business organisations are being affected and affect ecosystems. It provides a comprehensive framework for the business organisations for the management of natural resources on which their operations depend. The book concludes with an epilogue about designing sustainable ecosystems for equitable and sustainable development in the region.

This book has been written by leading researchers in South Asia. I believe that it will be useful for undergraduates, postgraduates, ecological economists, policy makers, local and regional government personnel, etc. I acknowledge the effort, knowledge and care of team members that brought this project to completion. I sincerely thank all co-authors for their valuable contributions and publishers for their encouragement to complete this book. I also thank my family for their continued support.

Adelaide, Australia Harpinder Sandhu
December 2016

Contents

Part I Setting the Scene

1 **Theory and Practice of Ecosystem Functions and Management** 3
Harpinder Sandhu

2 **Economic Systems and Ecosystems: Interlinkages, Co-evolution or Disparate Movement?** 17
Kanchan Chopra and Purnamita Dasgupta

Part II Ecosystems' Perspectives

3 **Interactions Between Humans and Ecosystems in Himalayas of India and Its Socioeconomic and Ecological Consequences: An Ecological Modelling Approach** 39
Sunil Nautiyal

4 **Ecosystem Services and Agriculture in Punjab, India** 59
Lalit Kumar and Harpinder Sandhu

5 **Desert Ecosystem Management: A Sustainable and Wise Use** 85
Aleem Ahmed Khan, Tahira Ruby, Nargis Naz and Muhammad Rafay

6 **Forest Ecosystem: Functions, Value and Management** 101
Madhu Verma, Rohit Singh and Dhaval Negandhi

7 **Urban Ecosystems: Functions, Value and Management** 123
Jürgen H. Breuste, Muhammad Mushahid Anwar, Rab Nawaz and Mehwish Rani

8 **The Hydro-Institutional Challenge of Managing Tanks: A Study of Tanks in Rural Andhra Pradesh** 155
M. Dinesh Kumar and Neena Rao

Part III Moving Forward

9 Interlinking Ecosystems for Multiple Benefits 195
Harpinder Sandhu

10 Sustainable Ecosystems and the Role of Business 207
Sukhbir Sandhu

Epilogue 219

Contributors

Muhammad Mushahid Anwar Department of Geography, University of Gujrat, Gujrat, Pakistan

Jürgen H. Breuste Department of Geology and Geography, University of Salzburg, Salzburg, Austria

Kanchan Chopra Formerly Institute of Economic Growth, University of Delhi Enclave, Delhi, India

Purnamita Dasgupta Formerly Institute of Economic Growth, University of Delhi Enclave, Delhi, India

Aleem Ahmed Khan Department of Zoology, Ghazi University, Dera Ghazi Khan, Pakistan

Lalit Kumar Department of Business Economics, Bhim Rao Ambedkar College, Delhi University, Delhi, India

M. Dinesh Kumar Institute for Resource Analysis and Policy, Hyderabad, India

Sunil Nautiyal Centre for Ecological Economics and Natural Resources (CEENR), Institute for Social and Economic Change (ISEC), Bangalore, India

Rab Nawaz Department of Geography, University of Gujrat, Gujrat, Pakistan

Nargis Naz Department of Life Sciences, The Islamia University of Bahawalpur, Bahawalpur, Pakistan

Dhaval Negandhi Indian Institute of Forest Management, Bhopal, Madhya Pradesh, India

Muhammad Rafay Department of Forestry, Range and Wildlife Management, The Islamia University of Bahawalpur, Bahawalpur, Pakistan

Mehwish Rani Department of Geography, University of Gujrat, Gujrat, Pakistan

Neena Rao SaciWATERs, Hyderabad, India

Tahira Ruby Department of Life Sciences, The Islamia University of Bahawalpur, Bahawalpur, Pakistan

Harpinder Sandhu School of the Environment, Flinders University, Adelaide, Australia

Sukhbir Sandhu School of Management, University of South Australia, Adelaide, Australia

Rohit Singh Indian Institute of Forest Management, Bhopal, Madhya Pradesh, India

Madhu Verma Indian Institute of Forest Management, Bhopal, Madhya Pradesh, India

Part I
Setting the Scene

Chapter 1
Theory and Practice of Ecosystem Functions and Management

Harpinder Sandhu

An ecosystem can be defined as a functional ecological unit comprising biophysical and chemical components that are interacting with each other (Lovelock 1979; Costanza et al. 1992; Gadgil and Guha 1993; Daily 1997). The science of ecosystems is the study of inter-relationships between these ecosystems their component and human beings. Ecosystems are broadly divided into two categories natural and managed (MEA 2005; Wratten et al. 2013). Natural ecosystems are the one with limited management by human beings or are not modified by humans, for example, forests, oceans, mountains etc. Whereas managed ecosystems are the ones, which are heavily modified by humans for their own use, for example agroecosystems (Sandhu et al. 2008). These ecosystems perform functions and provide a range of benefits in the form of ecosystem services (ES) for human well-being (Dasgupta 2001; MEA 2005). It is currently recognised that not only natural but also modified ecosystems perform certain functions, which significantly impact the delivery of ecosystem goods and services that contribute to human welfare (Daily 1997; Costanza et al. 1997, 2014; Wratten et al. 2013; Sandhu et al. 2015). In recent years, the concept of ES has gained wide acceptance within the international scientific community (MEA 2005). Extensive international peer review resulted in the adoption of the ES concept by the United Nations' sponsored Millennium Ecosystem Assessment (MEA) program (www.millenniumassessement.org). It was designed to meet the needs of decision makers and the public for scientific information concerning the consequences of ecosystem change for human well-being and options for responding to those changes. The MEA raised awareness of ecosystems and their services but it concluded that the global environment continues to degrade faster than ever before due to lack of understanding of the value of ES to the economy and well-being of humans. To fulfill this gap, recently, the United Nations has established the

H. Sandhu (✉)
School of the Environment, Flinders University, Adelaide, Australia
e-mail: Harpinder.Sandhu@flinders.edu.au

H. Sandhu (ed.), *Ecosystem Functions and Management*,
DOI 10.1007/978-3-319-53967-6_1

Intergovernmental Science-Policy Platform on Biodiversity and Ecosystem Services (IPBES 2010) to translate ecosystem science into action. This action plan is focused on strengthening assessment, relevant policy and associated science at a range of spatial and temporal scales for the globe. Moreover, there is a potential new paradigm shift at global level after the Rio+20 summit towards sustainable goals for decreasing poverty, hunger and improving environment sustainability (UN 2016). Therefore, understanding functioning of ecosystems and human reliance on them is critical to respond to the global goals of a desirable and sustainable future (Chopra and Kadekodi 1999; Kadekodi 2004; Kumar 2007, 2010). This chapter first describes the concept of ecosystem science by defining ES associated with various ecosystems. It provides global estimates of the economic value of the benefits and compare with those in South Asia. South Asia comprises seven countries: Afghanistan, Pakistan, India, Nepal, Bhutan, Bangladesh and Sri Lanka. This book largely focuses on this region and provides case studies citing examples from the key ecosystems. It concludes by providing policy and management options to protect these ecosystems for long-term sustainability and human well-being.

1.1 Types of Ecosystems

Ecologists have described various types of ecosystems based on their biophysical characteristics and their functions (Daily 1997; De Groot et al. 2002). We adopt the typology used by the MEA for consistency in this book. MEA divides global ecosystems into 10 broad categories: marine, coastal, inland water, forest, dryland, island, mountain, polar, cultivated and urban. These systems consist of several ecosystems that share similar biological, climatic, and social factors. These are briefly described in Table 1.1.

In this book, we explore six different types of ecosystems and provide case studies that examine the ecological and socio-economic dimensions of these ecosystems located in South Asia (Negi 1993; MEA 2005). Six key ecosystems that cover most of this region are briefly defined below.

1.1.1 Mountain Ecosystems

Mountain ecosystems consist of areas with elevation between 300 and 1000 m with slope of at least 2° (Körner and Ohsawa 2005). Globally about half of the human population depends on mountain ecosystem for various benefits especially water (MEA 2005; ICIMOD 2015). The dominant mountain region encompassing South Asia is the Himalayan rages extending 3500 km from Afghanistan in the west to Myanmar in the east (Sandhu and Sandhu 2014). More than 1.2 billion people live in mountain ecosystems. There is significant number of population in South Asia that depends directly on mountain ecosystems especially in the Himalayas.

Table 1.1 Description of ecosystems adapted from the Millennium Ecosystem Assessment (adapted from MEA 2005)

Ecosystems	Description
Marine	Ocean, with fishing. Marine areas where the sea is deeper than 50 m
Coastal	Interface between ocean and land, extending seawards to about the middle of the continental shelf and inland. Area between 50 m below mean sea level and 50 m above the high tide level or extending landward to a distance 100 km from shore. Includes coral reefs, intertidal zones, estuaries, coastal aquaculture, and seagrass communities
Inland water	Permanent water bodies inland from the coastal zone. Rivers, lakes, floodplains, reservoirs, and wetlands; includes inland saline systems
Forest	Land dominated by trees. A canopy cover of at least 40% by woody plants taller than 5 m
Dryland	Lands where plant production is limited by water availability. Annual precipitation is less than two thirds of potential evaporation, from dry sub humid areas
Island	Lands isolated by surrounding water
Mountain	Steep and highlands. Specifically, elevation >2500 m, elevation 1500–2500 m and slope >2°, elevation 1000–1500 m and slope >5° or local elevation range (7 km radius) >300 m, elevation 300–1000 m and local elevation range (7 km radius) >300 m, isolated inner basins and plateaus less than 25 km^2 extent that are surrounded by mountains
Polar	High latitude systems frozen for most of the year. Includes ice caps, areas underlain by permafrost, tundra, polar deserts, and polar coastal areas
Cultivated	Lands dominated by domesticated plants and animals. Areas in which at least 30% of the landscape comes under cultivation in any particular year
Urban	Built environment with a high human density. Known human settlements with a population of 5000 or more

1.1.2 Agroecosystems

Agroecosystems comprises of rural land that support activities such as farming and rural residential. Agriculture is one of the largest sector of global economy that supports 45% of the global population as farmers, labours, agribusiness organisations and contribute to the provision of ecosystem goods and services (Wratten et al. 2013). It contributes on an average 6% to the global gross domestic product (GDP), ranging from only 1% to that of advanced economies to 40% in least developed economies (World Bank 2016). Agroecosystems occupy about 5 billion hectare (one-third) of the global land area, and are the largest managed ecosystems on earth. Out of these 5 billion, about 1.5 billion hectare is arable (cropping) land and rest is under pastures. In South Asia, key countries occupy over 400 million hectares with little over 200 million hectares in agriculture (FAOSTAT 2016).

1.1.3 Desert Ecosystems

Desert ecosystems are the ecosystems that have barren land and receive very less precipitation (Safriel and Adeel 2005; Sivaperuman et al. 2009). Hence, have very sparse plants, animal and human population. Globally deserts occupy about one-third of the land surface. In South Asia, they occupy about 30 million hectares mostly located in the eastern parts of Pakistan and western parts of India (FAOSTAT 2016).

1.1.4 Forest Ecosystems

Forest comprises of natural flora and fauna that interacts with physical features of the environment (Shvidenko et al. 2005). Forest ecosystems provide several benefits to human beings in terms of timber, wild foods, and biological resources to be used in pharmaceuticals, provision of freshwater and clean air, biodiversity etc. In South Asia forest occupy about 80 million hectares which is approx. 20% of the total land area (FAOSTAT 2016).

1.1.5 Urban Ecosystems

Urban ecosystems comprise of areas which are built area for human habitat (Breuste et al. 2013). These are not natural areas but have been modified heavily to provide infrastructure to host human population. However, they depend heavily on surrounding ecosystems for all goods and services. For example, rural ecosystems for food, forests for biodiversity and fresh air, rivers for freshwater etc. Recent data estimates that 50% of the human population lives in cities and towns which was less that 20% at the beginning of 20th century (UN 2015). It is estimated that more than 75% population will move to urban areas in next 40 years. In South Asia, out of total 1.5 billion population, 30% lives in rural areas (FAOSTAT 2016).

1.1.6 Freshwater Ecosystems

Inland rivers, reservoirs, ponds, groundwater resource comprise freshwater ecosystem that sustain human, plant and animal life on the earth. Out of the total water available only 2.5–3.0% is freshwater and 1% of it is available for human use whereas rest is stored in ice caps, groundwater etc. (Vörösmarty et al. 2005).

1.2 Benefits from Ecosystems

Human beings are integral part of global ecosystems and they receive benefits in the form of food, timber, freshwater etc. from these ecosystems (Costanza et al. 1997; De Groot et al. 2002; Sandhu et al. 2008, 2012; Wratten et al. 2013). These benefits are termed as the ecosystem services (Table 1.2; Daily 1997). These ES in the form of material and non-material benefits results in human well-being (Costanza 2008). The linkages between ecosystems, its services and human well-being are classified as an ecosystem based approach (UNEP 2010). This approach is based on the principles of equity and sustainability for the integrated management of land, water, and living resources on the planet. For implementing this approach there is need to understand the complexities between benefits of ecosystems and relationship with human well-being. Therefore, key benefits in terms of ES provided by six ecosystems in South Asia are described below and in Table 1.3. Based on MEA, these ES are divided into four categories: provisioning, regulating, cultural and supporting services. Provisioning services are the outputs of regulating and supporting services and are primarily used for human consumption. For example food, fruits, timber are the key ES in this category. Regulating services are defined as the processes that regulate key ecological process such as climate regulation by storing carbon in soil or vegetation. Cultural services are the non-material benefits such recreation, aesthetics etc. Supporting services like pollination, nutrient cycling support production of provisioning services.

1.2.1 Mountain Ecosystems

Mountains provide several regulating services (such as clean air, flood protection) cultural services (aesthetics, source of inspiration, recreation etc.) but are limited in the supply of provisioning and supporting services.

1.2.2 Agroecosystems

Agroecosystems are designed to provide provisioning services in the form of crops, horticultural products, fibre, meat, etc. Regulating and supporting services are limited in supply from these ecosystems. Some agroecosystems also provide opportunities for cultural ES.

Table 1.2 Classification of ecosystem services (Costanza et al. 1997; De Groot et al. 2002; MEA 2005; Sandhu et al. 2012; Wratten et al. 2013)

	Ecosystem services	Definition	Example
Provisioning services			
1	Food production	That portion of gross primary production extractable as food	Production of fish, crops, nuts, fruits
2	Raw material	That portion of gross primary production extractable as raw material	Production of timber, fuelwood, or fodder
3	Genetic resources	Sources of unique biological materials and products	Products for materials science, resistance to plant pathogens and crop pests
4	Ornamental resources	For display purpose	Horticultural products, flowers etc.
5	Medicinal resources	Source of medicinal compounds	Products used in medicines
Regulating services			
6	Gas regulation	Regulation of atmospheric chemical composition	CO_2/O_2 balance, O_2 for UVB, SO_x levels
7	Climate regulation	Regulation of global temperature, precipitation, and other biologically mediated climatic processes at global or local levels	Greenhouse gas regulation
8	Disturbance regulation	Capacitance, damping and integrity of ecosystem response to environmental fluctuations	Storm protection, flood control, drought recovery
9	Water regulation	Regulation of hydrological flow	Irrigation, milling transportation
10	Water supply	Storage and retention of water	Watersheds, reservoirs, aquifers
11	Erosion control and sediment retention	Retention of soil within an ecosystem	Erosion control, reduction of runoff
12	Waste treatment	Recovery of mobile nutrients and removal or breakdown of excess or xenic nutrients and compounds	Waste treatment, pollution control detoxification
13	Refugia	Habitat for resident and transient production	Nurseries, habitat for migratory species, regional habitats for locally harvested species
Cultural services			
14	Aesthetic information	Associated landscapes	Landscaping of farmland
15	Recreation	Providing opportunities for recreational activities	Eco-tourism, sport fishing, outdoor activities

(continued)

Table 1.2 (continued)

	Ecosystem services	Definition	Example
16	Cultural and artistic information	Providing opportunities for non-commercial uses	Aesthetic, artistic, education spiritual, and/or scientific values,
17	Spiritual and historic information	Source of historic and spiritual value	Associated history of farmsteads
18	Science and education information	Source of education and training	Research and development
Supporting services			
19	Pollination	Movement of floral gametes	Reproduction of plant populations
20	Biological control	Trophic-dynamic regulations of population	Reduction of herbivory by top predators, control of prey species
21	Carbon accumulation	Carbon sequestration by vegetation and soil	Regulation of chemical composition
22	Mineralisation of plant nutrients	Storage, internal cycling, processing and acquisition of nutrients	Nitrogen fixation
23	Soil formation (maintenance of soil health)	Soil formation processes (Turning over of soil by earthworms	Structure maintenance
24	Nitrogen fixation	Storage and cycling	Legumes fixing N
25	Services provided by shelterbelts	Protection against wind erosion	Windbreaks

Table 1.3 Ecosystem services and their intensity (XXX = high, XX = medium, X = low) in six key ecosystems in South Asia

Ecosystems	Provisioning	Regulating	Cultural	Supporting
Mountain		XXX	XX	
Agriculture	XXX		X	XXX
Desert	X	X	X	
Forest	X	XXX	XX	X
Freshwater	XXX		XX	
Urban		XX	XX	

1.2.3 Desert Ecosystems

Desert ecosystems are also limited by biophysical constraints to supply ES in general. These ecosystems provide habitat to arid fauna and flora and provide provisioning services in the form of dry land agriculture products. Deserts also provide some opportunities for tourism.

1.2.4 Forest Ecosystems

Forests are the most valuable source of regulating services. They provide habitat to wild species of plants and animals. Vegetation in trees and shrubs capture and store carbon which is a key pollutant driving climate change. Forests also provide cultural services such as recreation and aesthetics, source of spiritual aspirations etc.

1.2.5 Urban Ecosystems

Urban ecosystems host more than half of the human population. They do not provide provisioning services but are source of cultural services. Urban ecosystems are more reliant on other ecosystems for the fulfillment of needs. However, they generate economic activities which result in human welfare.

1.2.6 Freshwater Ecosystems

Inland sources of freshwater such as reservoirs, lakes, ponds, and rivers provide essential provision services in the form of freshwater, fisheries etc.

1.3 Valuation of Ecosystems

Decision making at local, national and global level very often ignore values associated with ecosystems and their services (UNEP 2010; UN 2013). This happens when these benefits are unaccounted for in the national income or the gross domestic product (GDP; Sen et al. 2010; Kubiszewski et al. 2013; NSO 2013). The utilitarian approach argues that there are benefits from ES and hence we can attach economic values (Costanza 2008). These values could be direct use value of ES or indirect use value (De Groot et al. 2002). More on economic valuation is presented in the chapter nine in this book. Valuing ecosystems and their services provide an estimate of the natural capital available in the region. This assessment and valuation is important tool that can be used for the protection of those natural assets.

Global ES have been estimated to be in the range of US$125–145 trillion annually (Costanza et al. 2014). It includes value of 16 ES in major biomes of the world. According to an estimate by Costanza et al. (2014), global loss of ES is also on increase at US$20 million annually. This is partly due to lack of integration of ES approach in managing ecosystems and also due to lack of understanding of their role in generating human welfare. By using similar methodology of benefit transfer used by Costanza et al. (2014), economic values of four categories of ES associated with three key ecosystems in six major countries (Bangladesh, Bhutan, India, Nepal, Pakistan and Sri Lanka) are presented in Table 1.4 (in 2016 US$). These values are then used to estimate the total economic value of three key ecosystems according to their geographical area (Table 1.5). This snapshot of valuation indicates that the agroecosystems are the main providers of provisioning and supporting services whereas regulating services are highest in forest ecosystems. Urban ecosystems are the main providers of cultural services. These estimates are based on the studies used to estimate the total economic values by Costanza et al. (1997, 2014). These estimates are used here to provide a magnitude of the values associated with ecosystems in South Asia. For comparisons, Table 1.6 summarises total GDP (World Bank 2016), share of agriculture and forest rent in six countries in south Asia. Agroecosystems in these countries provide $1447 billion worth of annual (2.7 times more) benefits or ecosystem services than the current agriculture share ($532 billion) in GDP. Similarly, forest ecosystems also provide $355 billion worth of annual benefits in terms of ecosystem services (7.7 times more) than the forest share ($46 billion) in GDP. Total benefits worth $2800 billion annually

Table 1.4 Economic values (US$ per ha per year) of four categories of ecosystem services in three major ecosystems in South Asia

Ecosystems	Provisioning services	Regulating services	Cultural services	Supporting services	Total economic value
Agroecosystems	4127	1514	94	770	6505
Forest ecosystems	1002	1972	1099	359	4432
Urban ecosystems	0	1060	6610	0	7670

Table 1.5 Area (million ha) under three key ecosystems in six countries and their total economic value (US$ billion per year) in South Asia

	Agroecosystems		Forest ecosystems		Urban ecosystems	
	Area (M ha)	Total economic value (US$ billion per year)	Area (M ha)	Total economic value (US$ billion per year)	Area (M ha)	Total economic value (US$ billion per year)
Bangladesh	9.30	60	1.44	6.4	3.03	23
Bhutan	0.56	3.6	3.25	14	1.39	11
India	179.71	1169	68.43	303	92.66	711
Nepal	4.21	27	3.64	16	2.45	19
Pakistan	26.20	170	1.69	7.5	29.28	225
Sri Lanka	2.64	17	1.86	8	1.13	9
Total	223	1447	80	355	130	998

Table 1.6 GDP, agriculture share and forest rent in six major countries in South Asia

	GDP US$ billion per year	Agriculture (% of GDP) US$ billion per year	Forest rents (% of GDP) US$ billion per year
Bangladesh	195	31 (16%)	3 (1.5%)
Bhutan	2	0.4 (18%)	0.5 (25.1%)
India	2073	352 (17%)	41 (1.8%)
Nepal	21	7 (34%)	0.15 (7.1%)
Pakistan	270	67 (25%)	0.27 (1.3%)
Sri Lanka	82	74 (9%)	0.6 (0.7%)
Total	2643	532 (20%)	46 (1.7%)

provided by only three ecosystems—agriculture, forest and urban—out performs the combined GDP of $2642 billion in this region.

Valuation of ecosystems and their services can provide necessary information about the significance of ecosystems for human well-being and to formulate policy for their protection. Very often valuation is misinterpreted as an exercise aimed at commodifying nature (Costanza et al. 2014). However, on the contrary, valuation helps to identify various trade-offs between ES in particular ecosystems. As ES are associated with public or common goods, trading them in conventional market will not work properly (Costanza et al. 2014). Valuation approach includes use or non-use values of ES and not the exchange values. Therefore, valuation of ecosystems should not be confused with privatization of nature.

1.4 Ecosystem Management

Ecosystems provide both economic and non-economic benefits to humans through various ES. However, they cannot provide any benefit without human intervention. Therefore we emphasise on the human aspects of ecosystem management in this book. This notion of ecosystem management is also aligned with the earlier studies such as MEA and the current United Nations led Intergovernmental Science-Policy Platform on Biodiversity and Ecosystem Services (IPBES). The conceptual framework adapted in this book builds on earlier works where main elements are human well-being which is dependent on the interaction of various biotic and abiotic factors leading to the generation of ES (Fig. 1.1). It also accepts that human activities lead to various impacts on ecosystems and their services through socio-economic and biophysical drivers. Ecosystem based management is required at temporal and spatial scales to maintain ecosystems and sustain human welfare.

The framework of ecosystem based management presented here includes social, economic and natural capital for human well-being. It is widely accepted by the international scientific community (Costanza et al. 1997; Daily 1997; Sandhu et al. 2008, 2012, 2015, 2016; Wratten et al. 2013). It can be used effectively by policy makers to resolve some of the issues related with environmental degradation and

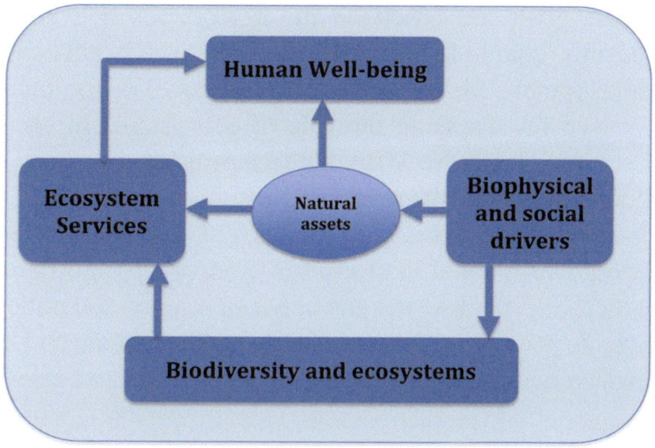

Fig. 1.1 Conceptual framework of human well-being supported by biodiversity and ecosystems

alleviation of poverty through global efforts through the recently set up sustainable development goals for the globe to target extreme poverty, hunger, child nutrition, gender issues, environment etc. (UN 2015). ES framework has been applied widely for environmental conservation, human well-being studies and for sustainable development interventions for poverty alleviation. Therefore, the need is to develop pathways based on the broad framework of ES to achieve equitable and sustainable development (Sandhu and Wratten 2013; Sandhu and Sandhu 2015).

Scientific evidence suggests that human activities have been impacting global ecosystems and their services since industrialisation (IPCC 2014). Recent developments also indicate that these ecosystems support human life on the planet through various processes and functions. To manage natural resources for the benefit of human society, the United Nations has started environmental programs that include policies and practices for the protection of ecosystems. These policies form the basis of ecosystem management practices for the various ecosystems around the globe. These are used to inform governments, scientific community, private sector, and the general public to develop responses for the sustainable management of ecosystems. Ecosystem management approach has its origin in the United Nations Environment Program (UNEP 2010) led environmental movement since 1970 through the various action programs to understand and internalise the role of environment in socio-economic development policies. Beginning in Rio Summit of 1992, UNEP led the process of raising concern and awareness of environment degradation on the planet and its impact s on human health and economy. These principles became the basis of the Millennium Development Goals (MDGs) and the Millennium Ecosystem Assessment (MEA) in beginning of 21st century. MDGs were specific goals aimed at reducing poverty, hunger and environmental degradation, dealing with child malnutrition, maternal health and education issues across all member countries of the United Nations (UN 2000). MDGs were initially setup for the period between 2000 and 2015. As these goals came to end in 2015, UN has setup Sustainable Development Goals (SDGs; UN 2015) for the post 2015 development agenda (https://sustainabledevelopment.un.org/topics/sustainabledevelopmentgoals).

These goals are based on the agreements between UN member countries during recently concluded Rio+20 Summit, which advocates a sustainable and equitable development for all in the 21st century. For the management of ecosystems, MEA focused on assessing the role of ecosystems in socio and economic development. Learnings from the MEA led to a program of action through the global study—The Economics of Ecosystems and Biodiversity (TEEB 2010). Now to translate this science of ecosystems into practice, UN have started an action plan through the Intergovernmental Platform for Biodiversity and Ecosystem Services (IPBES 2010), which aims to close the gap between science and policy at global and regional level. It aims to provide action plan for the governments and society to implement ecosystem management in their development programs to achieve the agenda 2030 (UN 2016).

1.5 Summary

Natural and managed ecosystems provide many benefits for human and industrial use to fulfill basic needs and to support global economy. The unprecedented growth in human activities since industrialisation started in 18th century have led to massive environmental degradation worldwide. Scientific, policy and civil community has raised concerns about these unwanted developments, which threaten the existence of human beings on the planet. Various global and regional programs under the United Nations have raised awareness about the issues and generated policy responses to deal with those issues at local, regional or global scale. Scientific and policy community along with the civil society are working in collaboration to address these global issues for positive outcomes for all species of the planet (UNDP 2013). This response also recognises the high economic value of ecosystems and their services to local and global economy. Rest of this book elaborates on this framework and provides key concepts about the role of ecosystem management in sustaining human well-being and the economy by focusing on South Asia. It also explores various case studies by taking examples from different ecosystems and examine their role in equitable and sustainable development in South Asian region.

References

Breuste J, Haase D, Elmquist T (2013) Urban landscapes and ecosystem services. In: Wratten S, Sandhu H, Cullen R, Costanza R (eds) Ecosystem services in agricultural and urban landscapes. Wiley Blackwell, Oxford, pp 83–104

Chopra K, Kadekodi G (1999) Operationalising sustainable development: economic-ecological modelling for developing countries. SAGE Publications Pvt. Ltd, New Delhi

Costanza R (2008) Ecosystem services: multiple classification systems are needed. Biol Conserv 141:350–352

Costanza R, Norton B, Haskell BJ (eds) (1992) Ecosystem health: new goals for environmental management. Island Press, Washington, DC pp 269
Costanza R, d'Arge R, de Groot R, Farber S, Grasso M, Hannon B, Limburg K, Naeem S, O'Neill RV, Paruelo J, Raskin RG, Sutton P, vans den Belt M (1997) The value of the world's ecosystem services and natural capital. Nature 387:253–260
Costanza R, de Groot R, Sutton P, van der Ploeg S, Anderson SJ, Kubiszewski I, Farber S, Turner RK (2014) Changes in the global value of ecosystem services. Glob Environ Change 26:152–158
Daily GC (ed) (1997) Nature's services: societal dependence on natural ecosystems. Island Press, Washington, DC
Dasgupta P (2001) Human well-being and the natural environment. Oxford University Press, Oxford
De Groot RS, Wilson M, Boumans R (2002) A typology for the description, classification and valuation of ecosystem functions goods services. Ecol Econ 41:393–408
FAOSTAT (2016) Food and Agriculture Organisation of the United Nations. http://faostat.fao.org/
Gadgil M, Guha R (1993) This Fissured Land: an ecological history of India. University of California Press, California
ICIMOD. 2015. International Centre for Integrated Mountain Development, Nepal. http://www.icimod.org/
IPBES (2010) Intergovernmental science-policy platform on biodiversity and ecosystem services. Analysis of the assessment landscape for biodiversity and ecosystem services. UNEP, Busan, Republic of Korea, 7–11 June 2010. http://www.ipbes.net/sites/default/files/downloads/UNEP_IPBES_3_INF_1_Add.1_EN.pdf
IPCC (2014) Climate change 2014: summary for policymakers. Synthesis report: Geneva. http://www.ipcc.ch/report/ar5/
Kadekodi G (2004) Common property resource management: reflections on theory and the Indian experience. Oxford University Press, Oxford, UK
Körner C, Ohsawa M (2005) Mountain systems. In: Current state & trends assessment, Millennium ecosystem assessment. Island Press, Washington. pp 681–716
Kubiszewski I, Costanza R, Franco C, Lawn P, Talberth J, Jackson T, Aylmer C (2013) Beyond GDP: measuring and achieving global genuine progress. Ecol Econ 93:57–68
Kumar P (2007) Economics of environment and development. CRC Press
Kumar P (2010) The economics of ecosystems and biodiversity: ecological and economic foundations. TEEB—The Economics of Ecosystems and Biodiversity, Routledge
Lovelock J (1979) Gaia: a new look at life on earth. Oxford University Press, Oxford
MEA (2005) Millennium ecosystem assessment synthesis report. Island Press, Washington
National Statistical Organization (2013) Green national accounts in India: a framework. Ministry of Statistics and Programme Implementation Government of India, p 256
Negi SS (1993) Biodiversity and its conservation in India. Indus Publishing, New Delhi
Safriel U, Adeel Z (2005) Dryland systems. Current state & trends assessment, Millennium ecosystem assessment. Island Press, Washington, pp 623–662
Sandhu HS, Wratten SD, Cullen R, Case B (2008) The future of farming: The value of ecosystem services in conventional and organic arable land. An experimental approach. Ecol Econ 64:835–848
Sandhu H, Sandhu S (2014) Linking ecosystem services with the constituents of human well-being for poverty alleviation in eastern Himalayas. Ecol Econ 107:65–75
Sandhu H, Sandhu S (2015) Poverty, development, and Himalayan ecosystems. Ambio 44:297–307
Sandhu H, Crossman N, Smith F (2012) Ecosystem services and Australian agricultural enterprises. Ecol Econ 74:19–26
Sandhu H, Wratten S (2013) Ecosystem services in farmland and cities. In: Wratten S, Sandhu H, Cullen R, Costanza R (eds) Ecosystem services in agricultural and urban landscapes. Wiley Blackwell, Oxford

Sandhu H, Wratten S, Costanza R, Pretty J, Porter J, Reganold J (2015) Global significance of non-traded ecosystem services on farmland. PeerJ 3:e762. doi:10.7717/peerj.762
Sandhu HS, Wratten SD, Porter JR, Costanza R, Pretty J, Reganold J (2016) Mainstreaming ecosystem services into future farming. Solutions 7:40–47
Sen A, Stiglitz J, Fitoussi JP (2010) Mismeasuring our lives: why GDP doesn't add up: the report. New Press Distributed by Perseus Distribution, New York
Shvidenko A, Barber CV, Persson R (2005) Forest and woodland systems. Current state & trends assessment, Millennium ecosystem assessment. Island Press, Washington, pp 585–621
Sivaperuman C, Baqri QH, Ramaswamy G, Naseema M (eds) (2009) Faunal ecology and conservation of the great Indian desert. Springer, Berlin
TEEB (2010) The economics of ecosystems and biodiversity: mainstreaming the economics of nature: a synthesis of the approach, conclusions and recommendations of TEEB
United Nations (2000) Millennium development goals. http://www.un.org/millenniumgoals/
United Nations (2015) Sustainable development goals. https://sustainabledevelopment.un.org
United Nations (2013) The system of environmental—economic accounting (SEEA). Department of Economic and Social Affairs Statistics Division, United Nations
United Nations (2016) United Nations population division. http://www.un.org/en/development/desa/population/
United Nations Environment Program (2010) The Role of ecosystems in developing a sustainable 'Green Economy'. UNEP Policy Series, Brief 2
United Nations Development Program (2013) The rise of the South: human progress in a diverse world. Human Development Report 2013. UNDP, New York
Vörösmarty CJ, Lévêque C, Revenga C (2005) Freshwater. In: Current state & trends assessment, Millennium ecosystem assessment. Island Press, Washington, pp 165–207
World Bank (2016) Gross domestic product (GDP). http://data.worldbank.org/indicator
Wratten S, Sandhu H, Cullen R, Costanza R (eds) (2013) Ecosystem services in agricultural and urban landscapes. Wiley-Blackwell, Oxford

Chapter 2
Economic Systems and Ecosystems: Interlinkages, Co-evolution or Disparate Movement?

Kanchan Chopra and Purnamita Dasgupta

2.1 Economic Systems and Ecosystems: How Do They Relate to Each Other?

In general parlance, an economic system is a fairly well-understood term, while 'eco-systems' is a somewhat esoteric one. An economic system refers to the production, distribution and consumption of goods and services in an economy and the network of societal arrangements and institutions governing them. Ecosystems are nature based categories. They have been defined as "a dynamic complex of plant, animal, and microorganism communities, and the non-living environment interacting as a functional unit." In other words, the natural world as we understand it can be thought of as consisting of a range of ecosystems.[1] One would expect that the manner in which humankind organizes its economic activity or the ways in which it provides for its material provisioning should be based on these subdivisions of the natural landscape. This happens only in a very limited manner. For the most part economic systems and the buzz that surrounds them overshadows and supersedes substantially the role of nature. We specify below how this happens.

Economic systems are categorized, more often than not, with reference to the significant institutions through which their functions of providing for human con-

[1]The Millennium Assessment (MEA 2005) identified nine major ecosystem types among them forests, fresh water, marine, agricultural and drylands.

K. Chopra · P. Dasgupta (✉)
Formerly Institute of Economic Growth, University of Delhi Enclave, Delhi, India
e-mail: pdg@iegindia.org

K. Chopra
e-mail: choprakanchan14@gmail.com

H. Sandhu (ed.), *Ecosystem Functions and Management*,
DOI 10.1007/978-3-319-53967-6_2

sumption are performed. Broadly, three kinds of economic systems are distinguished between: traditional, centrally planned and market, respectively those where the central organizing institution is tradition, central government planning (or command) and the market. Given that a range of traditional and government institutions, market structures and social customs are inevitably associated with all economic activity, this means that each of the three economic systems referred to above focuses more on some aspects to the neglect of others. Moreover, there has occurred a gradual transition from systems based on tradition towards the other two, even as it may sometimes be claimed that all three continue to co-exist simultaneously.

Another perplexing fact is that market dominated economic systems capture a great deal of attention in the discipline, even though historically speaking, they have had a relatively short time span of existence. People were hunter-gatherers for 99.5% of human history, until the rise of agriculture began some 15,000 years ago, and markets became significant much later. Even so, production, consumption and distribution of goods and services, by markets of different kinds seem to occupy the attention of economists most of the time. This may be due to the oft-mentioned fact that whereas the rules designed by societies or by central fiat need norms and laws to be implemented, markets provide, "self-correcting" systems where the forces of demand and supply working through prices (whether of goods, services or factors of production) ensure course-correction. It was postulated that economic activity took place when factors of production entered into contracts in market situations.

The reality however is that some aspects of production, consumption and social welfare, in all economies, lie outside markets. In the main, ecosystems and the services they provide as a category belong here. For a large part of human history, when the scale of economic activity was localized and limited, these systems constituted the main focal point of economic activity. However, technological change and the ever increasing scale of production and consumption expanded the reach and operating capacity of human creativity. The organization of economic activity took place in ways far removed from the natural resource base. As the Millennium Assessment put it, this resulted in increased human well-being for large populations but also led to a lacuna in the formal understanding of the association between natural and economic systems.

2.1.1 Are Ecosystems an Intrusion into the Functioning of Economic Systems? or, Is It the Other Way Round?

Currently, a large part of the literature in the context of nature and society is about re-conceptualising the links between nature and economic systems and their significance under different situations. One way to look at the role of nature is as an intrusion into the economic space, with nature limiting the space through constraints imposed by ecosystems which in turn become binding on economic activity. This is the widely held understanding of ecological issues limiting economic activity. It is a variation of the traditional 'externality' argument in economics. In this mode of thinking, human well-being in the present is of the essence. Anything which limits an extension of the economic system to enable it to achieve higher levels of current human welfare is considered an intrusion.

A more in-depth awareness of the science behind ecosystem state and functioning results in nuances in this approach bringing in the role of irreversible as against reversible change in ecosystem states and of discontinuous changes in ecosystems. These impact economic welfare through their impact on the scale of economic activity. Witness the literature on the ecology of shallow lakes and the consequences for their economic management (Carpenter 2001; Scheffer 1997): *the deep economics of shallow lakes, as stated by an experienced practitioner of the art.*[2] Economic systems and ecosystems are treated as *interlinked entities* in this mode of analysis.

Alternative approaches, formulated under the rubric of 'ecological economics', have taken a position that the *economic system is a part of a larger ecosystem*, conceived of as the sum total of the physical world with the laws of thermo-dynamics governing it. This larger viewpoint generated perhaps a scientifically more correct view of the world (such as that of a 'space ship earth' perspective and so on). It ordained that the safe operating space within which humanity could function was determined by planetary boundaries of different kinds.[3]

But it had two deficiencies. The first emerged from the fact that it took a very macro, physical view of the earth, which must manage its whole environment and give content to the notion of optimal physical scale of the economy relative to the ecosystem by minimizing throughput and energy use. But the planetary ecosystem that it conceived of is made up of several smaller ecosystems and the limits imposed are not equally binding on all sub ecosystems and at all times and places. Secondly, it left out of its reckoning the social and individual motivations for human behavior. Behavioral sciences are not integrated into the calculus of such ecological thinking. The ecological critique of economics has a fairly long lineage but has not been able to impact mainstream economics due partly to "the separation of the natural and the social sciences".[4] An important component in any integration shall have to be via the factors determining human behavior. Also, the unwillingness of these earlier forays to consider explicitly power relations in society has also been commented on.[5]

---

[2]See Maler (1999).

[3]One recent exposition of this is to be found in Rockstorm et al. (2009).

[4]See Martinez-Alier (1991).

[5]See for example, Gale and M'Gonigle (2000).

2.1.2 Both as Complex Adaptive Systems

We maintain in this chapter that economic systems and ecosystems can both be viewed as complex adaptive systems. In economies as well as in nature, individual agents compete for resources, and interact with one another directly through competition, exploitation or beneficial association or co-operation. In this interaction in the competitive economy, the coordination role of prices and markets was understood right from the time of Smith (1776) and culminated in the contributions of Leon Walras (1877) and Alfred Marshall (1890). This coordination worked both to ensure co-operation and to resolve conflict over resources. Ecological systems at all levels are similar to economic systems with possibilities of competition, co-operation and conflict between individual agents in the system. Both are "composed of individual agents which adjust their behavior or their relative numbers, with consequences for the system as a whole, and these consequences can in turn affect individual behaviours".[6]

Another similarity is that individual and community (or social in the economic context) goals do not often converge: incentive mechanisms may be inadequate: or may not exist at all. Further, both kinds of systems may collapse if thresholds are breached. Though economic systems exist within the larger biosphere or ecosystem, they ignore ecological thresholds in particular. This is often because market incentives do not exist to enable individual agents to internalize such irreversibilities. If anything it is the somewhat lower level of development of market institutions that prevents economic systems from taking ecosystem based risk and uncertainty into account. In this chapter we shall discuss two directions which can be taken to correct for this underlying deficiency of economic systems in their interaction with ecosystems.

The first direction is one in which independently generated scientific knowledge is used in the framework of risk analysis and management to set limits on the domain of economic systems. We argue that the links between ecosystem functioning and economic activity need to be looked at through the lens of risk, determining thereby, a safe operating space for humanity.

The second is one where economic decision-making is strengthened by attempting to put a value on hitherto unvalued ecosystem services, in particular of the regulating, supporting and cultural kinds. The notion of linking ecosystems to human well-being through the services they provide for humans is one of the plausible ways of placing nature and its contributions in the policy domain.

We conclude the chapter by focusing on ethical issues at the core of the interlinking between economic systems and ecosystems. The value we place on the present as against the future, (more mundanely, the social rate of discount) is fundamentally a human decision and is an ethical one. Since the time frame within which change occurs in the two kinds of systems differs greatly, it becomes of the

[6]See Arrow et al. in Barrett et al. edited (2014).

essence in determining the nature of the interaction. The economic system often drives the direction taken with a short-run perspective. It is here that the responsibility of humanity lies in adopting an informed and sufficiently long run perspective to ensure the continuance of a safe operating space.

2.2 Managing the Links Between Ecosystems and Economic Systems

2.2.1 *Risk Analysis and Management*

If nature is to be taken into account in the management of an economic system for development, the inherent uncertainties in the science of ecosystems should be known and factored in. We can then proceed with alternate frameworks of decision-making depending on the context for economic activity. Thus, where life-support services such as biodiversity or carbon sequestration are concerned, leaving decisions to the market will prove costly for long term economic growth. In others, such as recreational services and a range of provisioning services (food, fibre, fodder, fish) economic frameworks of analysis prove adequate where well-functioning markets exist, with built—in safe guards such as institutions to control perverse incentivisation so that basic principles provided by neoclassical economics are not violated. These include those such as the rate of regeneration not being lower than the rate of harvesting for renewable natural resources. However, here too, what is presumed is that the market for ecosystem services will function efficiently, at par with other product markets in terms of prices and costs in decision-making, with externalities fully accounted for. For instance, an assumption which is frequently violated in the real world is that of having no information asymmetry among agents. As Arrow et al. (2014) conclude, markets alone cannot create a society that is both just and sustainable.

It is, therefore, useful to adopt a ***risk analysis and management approach*** to prioritize amongst the ecosystem services which are instrumental to sustainable economic development, in particular when faced with the reality of tipping points and non-linearity in natural processes. Both the probability of and hence likely exposure to the occurrence of a change in an ecosystem service, and the economic system's ability to cope with these changes would determine the risks to social welfare (including change in physical or/and monetary values of economic assets, lives) and the need for foresight in planning for risk management. This is an important input into planning for investments in a manner conducive with acceptable levels of risk. For instance, a region well equipped with storm shelters and early warning systems may suffer low consequences, although it may fall in a high probability zone for cyclonic activity. Thus, an understanding of risks can result in appropriate planning, such that a location which is in a high-risk zone by mere biophysical parameterization, is transformed into a location in an acceptable risk zone for the community with judicious planning and investment.

An ecosystems approach towards economic decision-making has to focus on much beyond the biophysical as it seeks to accommodate the uncertainties of available knowledge, to minimize risks to ecosystems and communities dependent on these ecosystems. We consider two illustrations here that have relevance to the South Asian region, one in terms of rural areas where increasingly stress on water resources is being felt, partly due to climate change; and the other arising from pressures of fast paced urbanization.

Consider that climate change can lead to movement of species, leading to disruptions of ecosystem services such as water purification, pollination of crops or preservation of soils (Oppenheimer et al. 2014). Climate change adds to the stress caused by destruction of forested areas, acquisition of land for industrial activity and pollution of rivers, which in themselves have led to adverse impacts on biodiversity in many parts of South Asia. At the extreme, there is a risk of extinction of biodiversity. The resulting vulnerability is to the entire human and social systems as the loss in ecosystem services plays out in terms of challenges to food security, malnutrition and other such emergent risks that arise from the complex interactions of the socio-ecological and biological system. The implications for managing the risks arising from this vulnerability would be twofold:

(a) reduce the risk of biodiversity loss through non-market mechanisms such as use of a precautionary principle
(b) reduce the exposure and vulnerability of human systems through better management of land and water resources by using a range of policy instruments: market mechanisms such as efficient pricing of water to reflect its scarcity value and taxing water pollution; non market mechanisms such as protecting watersheds through legislation and community participation.

Fast paced urbanization has combined with the development of coastal areas in many parts of deltaic south Asia. These areas have become vulnerable to a host of stresses induced by sea level rise, coastal erosion, storm surges, flooding and saline intrusion (Oppenheimer et al. 2014). Climate change constitutes a threat multiplier, exacerbating the existing stresses in these areas and compounding of risks may take place as an initial risk leads to another cumulatively.

A risk management approach could comprise of:

(a) reducing impacts through adaptation strategies, building appropriate infrastructure (including through the creation of market based incentives)
(b) non market mechanisms such as investment by the public sector in projects that have high costs and low returns in the short run but high returns if all externalities are internalized (e.g. building storm shelters)
(c) managing the potential threats of displacement and conflicts among communities (those receiving populations and those displaced) through participatory and collective action among stakeholders. Thus, the impacts of extreme events can be managed, and is dependent on the economic options that are exercised to manage the risks (IPCC 2012, SREX).

These two illustrations highlight that (a) risks vary spatially and temporally and (b) effective risk management will to a large extent depend on the choices that are made based on value judgements about these risks depending on how different losses and damages are valued and ranked by society.

2.2.2 *Risk Levels and Risk Management*

Risk levels can be graded in terms of the probability of an adverse occurrence and the consequences of this occurrence. The extent of risk, whether high or low, will depend on the extent of exposure and vulnerability that the dynamic nature of the relationship between economic growth and the ecological system leads to. To quote from the IPCC glossary: risk is "the potential for consequences where something of human value (including humans themselves) is at stake and where the outcome is uncertain." In other words,

"Risk = (Probability of events or trends) × Consequences"

In the context of global warming for instance, the risks that may result from warming are graded from undetectable to very high, for various consequences in the recent IPCC report (IPCC, AR5, SPM WGII). Figure 2.1 presents how the risk level posed to human or natural ecosystem changes with the likely degree of warming (temperature changes discussed here are relative to 1986–2005). Current understanding is that the additional risk posed by warming at 1 °C to unique and threatened ecosystems has severe consequences, and can become very high at 2 °C warming. This includes the Arctic sea ice systems and coral reefs. However, the risks in terms of global aggregate economic impacts are moderate till 2 °C temperature rise, while risks from large scale singular events which comprise of

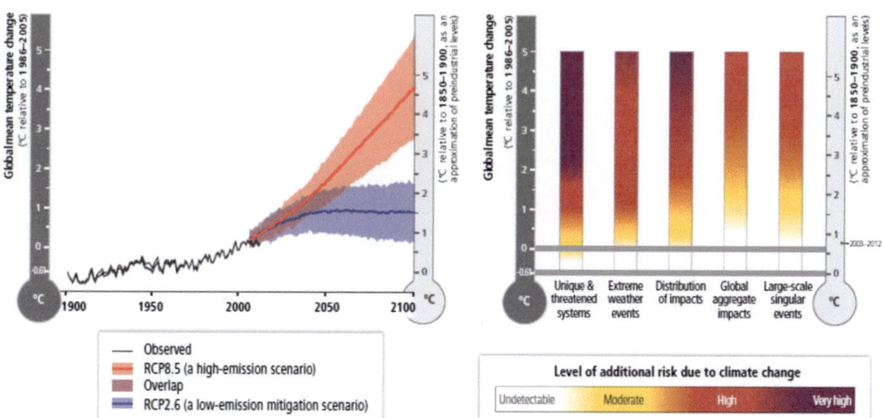

Fig. 2.1 Key risks across sectors and regions. *Source* Assessment Box SPM1. Fig. 1 (IPCC WG II, AR5, SPM, 2014). http://www.ipcc.ch/pdf/assessment-report/ar5/wg2/ar5_wgII_spm_en.pdf

irreversibilites of ecosystems are moderate between 1 and 2 °C, and rise disproportionately to become high risk as temperature increases from 1 to 2 °C. This would include large irreversible sea level rise due to loss of ice-sheet.

To understand the linkages across risks, the adverse occurrences in the natural world and their consequences, the risk to unique and threatened ecosystems, and global distributional impacts is further elaborated upon.

Risks to species can be represented in terms of the maximum speeds at which species can move across landscapes compared with speeds at which temperatures are projected to move across landscapes (Fig. 2.2). The white bar indicates the lower bound, median and upper bound of maximum speeds at which species can move across landscapes (compared with speeds at which temperatures are projected to move across landscapes). The left vertical axis measures speeds of movement, while the right vertical axis represents climate velocities for increasing temperatures. Horizontal lines show climate velocities for global land area averages (or for large flat regions) under alternative scenarios of warming for the future at 2.6, 4.5,

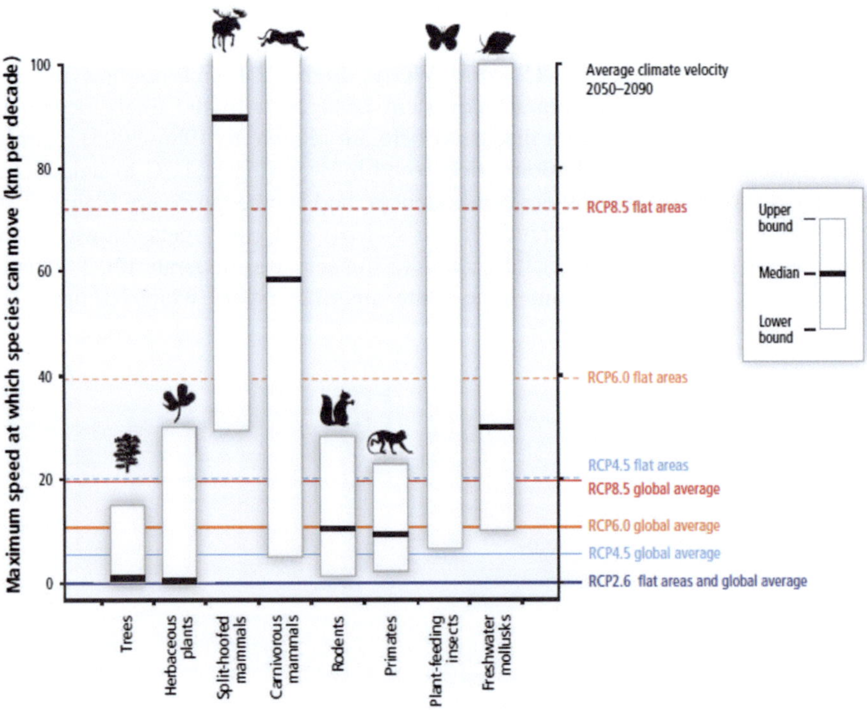

Fig. 2.2 Maximum speeds at which species can move across landscapes compared with speeds at which temperatures are projected to move across landscapes. *Source* IPCC, AR5, WG II, SPM: Fig. 5. http://www.ipcc.ch/pdf/assessment-report/ar5/wg2/ar5_wgII_spm_en.pdf

and so on futures (namely RCPs or representative concentration pathways). Species with maximum speeds that are below these lines, are unlikely to be able to move to overcome global warming without human intervention, and thereby likely to face extinction. Thus, primates are at risk and unlikely to survive at global warming corresponding to scenarios of 4.5 whereas plant feeding insects are not at risk for any of the scenarios shown.

Risk levels are impacted by factors that mitigate the probability of an adverse event's occurrence and or its consequences. For instance human intervention can influence the consequences of economic activity on the ecosystems. Consumption patterns, population and the availability of new technology, mediate this link between exposure levels of a system to a hazard, and its vulnerability.

Thus, economic policy instruments can become crucial in regulating and determining the extent of intrusion that occurs into natural systems, due to economic activity. However, as the very grouping of risks in Fig. 2.1 indicates, decision-making on the appropriate tools and instruments involves much more than a mere aggregation of the monetized damages arising from the threats posed by climate change. Rather it calls for a reasoned balancing of the overlaps and trade-offs among these risks, and the designing of interventions that are dependent on the scale, context and level of risk.

The argument can be elaborated upon, linking Figs. 2.1 and 2.2. Since climate change impacts crop production in different parts of the world differentially for the same global temperature change, the distributional impacts when aggregated may not be considered to be too high till a 2 °C warming level is breached. However, some ecosystems (for instance coral reefs) are already under risk, as are some unique ecosystems and cultures associated with them. The risk to these increases to a high level even with a 1 °C rise in temperature. Standard economic valuation techniques can provide a limited measure of the values that maybe lost due to extinction, given the current state of the art of these techniques. In recognition of these limitations, newer techniques such as non probabilistic methodologies and use of multi criteria analysis have been encouraged to supplement such measures in decision-making as to whether a 2 °C change in temperature would be acceptable or not in such circumstances.

A recognition of the interdependence between distributional impacts, species adaptive capacity and ecosystem well being, throws up the difficulty of using only estimates of aggregative economic damages for managing the threats. Risk management would include the use of a range of instruments that encompass the traditional, market and non-market societal values.

2.3 The Role of Valuation

The notion of value to humans is another significant way in which ecosystems have been linked to economic systems. This approach is rooted in economic science and is based on the assumption that services provided by ecosystems are anthropocentric in nature. As relative value is of the essence in making choices in economic theory, valuation is extended as a principle for factoring in ecosystem services.

It follows then that the role of monetary valuation of the services provided by ecosystems comes into its own as a guiding principle when trade-offs between different ecosystem services or with other goods and services is of the essence. On the other hand, risk analysis and management approaches are relevant when thresholds in ecosystem states are significant. Even after such thresholds have been factored in, there exist a range of ecosystem services for which tradeoffs are relevant.

2.3.1 Valuation of Ecosystem Services

It can be argued that ecosystems, through the services they provide, are of value to humans. Relative value, which leads to choice or trade-offs between services is often of the essence here. In other words, more availability of one service means less of the other. A choice has to be made. This choice does not become apparent if markets do not exist for both kinds of services. Typically, provisioning services from ecosystems have been privileged at the expense of regulating services because the former enter the metric of market valuation, while the latter do not. Although there are reasonable estimates of the value of many provisioning services—where markets are most well-developed—there are few reliable estimates of the value of most cultural services and all regulating services. The literature has, of course, thrown up a slew of methods for valuation of ecosystem services, including non market valuation. A discussion of these methods and their application to valuing ecosystem services can be found in Freeman (2003), and Pagiola et al. (2004). In recent years, the literature on valuation of ecosystem services has proliferated in particular with cases studies in different parts of the world using both revealed preference and stated preference methods.[7]

However, several aspects of the contribution made by ecosystem services to present and future human well-being are still largely not accounted for, among them the following:

- The effect which regulating services have on the distribution and supply of the provisioning and cultural services.

[7]See, among others, Haque et al. (2011).

- The exact nature of the links between the state of an ecosystem (such as forests) and the ecological services accruing there-from e.g. increasing forest cover may sometimes lead to decreases in water flows and flood regulation benefits may be lower than commonly assumed.
- The existence of ecological trade-offs and the disservices such as pathogens and pests which may accrue from some ecosystems.
- The future option values of some kinds of ecosystems.

Irrespective of reservations with respect to its accuracy and relevance, valuation of ecosystem services has found favour with policy makers, mainly because it fits into the economic money-metric. Such valuation, however incomplete or partial, is then used in the application of some methods for decision-making. A benefit-cost framework (BCA) is the most often used: more recent frameworks recommended are multi-criteria, deliberative methods or participatory methods. Decision makers almost always have a preference for relationships derived from empirical investigation which (provided they are expressed transparently)are more useful than those obtained only through deliberation[8] and BCA gets privileged on this count. This may result in ill-conceived analysis with erroneous policy implications, in particular when inputs from other approaches are not incorporated in decision-making.

In India, for instance, the understanding of the economic value of forest ecosystem goods and services in India is majorly impacted by various considerations depending on the scale at which the services are considered and the stakeholders' perceptions can vary significantly. On one hand, there is substantial mineral wealth below the ground in many forested regions of the country, for instance in states such as Orissa, Madhya Pradesh, Chattisgarh and Jharkhand. On the other, the forests themselves are of high and substantial ecological value, including biodiversity and habitat, and the sustenance of flows in rain-fed rivers which originate here and therefore support water related economic activities such as agriculture, downstream pastures, and inland navigation. Which of these should be given precedence? In this context, Mukhopadhyaya and Kadekodi (2012) provide an analysis of how a misplaced use of benefit-cost analysis with an under-valuation of forest ecosystem services can provide misleading pointers in favour of mining.

There is also a significant population of forest dwellers many of whom are critically dependent on these forests for both economic and socio-cultural reasons. The fact that large numbers of forest peoples (i.e. both forest inhabitants and forest adjacent communities) reside in states with poverty levels well above the national average, implies that decision-making based on the monetized values of ecosystem services would be erroneous in terms of its impact on sustainability of livelihoods of the poor. As per the methodology recommended by the Planning Commission Expert Group on Poverty (2009), the poverty level in states with a high number of forest dwellers is substantially higher. Concerns regarding the above issues have proved difficult to address in terms of quantitative approaches to deriving values. In

[8]See, for instance the discussion in Lele and Srinivasan (2013).

fact issues of marginalization and giving voice to diverse stakeholders, has led to the development of literature on the relevant language of valuation (Martinez-Alier et al. 2010) and on alternative approaches for assessing ecosystem services (Puroshothaman et al. 2013). At another level, recognition of the need to take "a middle of the road" approach have led to the refinement of methodology towards 'disaggregated benefit-cost approaches' as well (Lele and Srinivasan 2013).

Ecosystem services and their interdependencies have been assessed at macro scales as well. Climate change has forced recognition of the contribution ecosystem services make to economic growth at a macro level as increasingly evidenced by an upsurge in academic modelling exercises computing costs imposed by climate change in aggregate terms of loss in GDP (Stern 2007; IPCC 2014). At a macro scale, for instance, ADB and IFPRI (2009) estimate that for Asia, the annual spending for coping with adverse agricultural impacts of climate change between 2010 and 2050 lies in the range of \$4.2–\$5 billion. At the sectoral level, agriculture has received attention in costing and valuation studies on climate change. Dasgupta et al. (2013a) estimate that India could experience a decline of up to 18% in foodgrain production between 2030 and 2050. Kumar and Parikh (2001) estimated that a 2.0 °C rise in temperature and 7% increase in precipitation would lead to about 8.4% loss in net revenue in agriculture in India.

In other words, in a range of issues the determination of trade-offs, opportunity costs and benefits provides the metric for arriving at the value of different ecosystem services. However, whereas valuation has a role to play when trade-offs exist, it needs to be used in a very careful and context specific manner. Decision criteria that emerge from CBA or cost minimizing approaches are in themselves a reflection of the privileged position given to individual rationality, the dominant value underlying most of economic analysis. We need to further ask, whether this dominant value is also the correct one in analyzing the relation between ecosystems and the economic system. Could it be that the social constructivist position which assumes that an individual is 'socially constructed' may be a better starting point for analyzing the relationship between ecosystems and economic systems?

Further, and more importantly, in the valuation paradigm, we value ecosystem services as items (e.g. provisioning, regulating), with additive or non-additive properties, whereas they can be viewed only partly as items as they are also intrinsic outputs of ecosystem structures. Vatn (2009) illustrates this aptly with reference to Wilson's definition of biodiversity.[9] This definition captures two different traditions of defining biodiversity, firstly as a set of species and secondly as a system feature.

Does valuation capture both? Very unlikely; hence the limits of this approach in examining the relationship of economic and ecosystems and the imperative to bring

[9]Wilson (2001) defines biodiversity as "The variety of organisms considered at all levels, from genetic variants belonging to the same species through arrays of species, to arrays of genera, families, and still higher taxonomic levels; includes the variety of ecosystems, which comprise both the communities of organisms with particular habitats and the physical conditions under which they live" (p. 377).

in an analysis of system structures, and likelihoods of changes in them through a reliance on risk analysis and management.

2.3.2 Can Valuation Play a Role in Risk Management Approaches?

The role that economic valuation can play in building ecological sustainability, can also be visualised in terms of its contributions towards decision-making that reduces vulnerability to risks and increases resilience of both human and natural systems. In recent years, there has increasingly been a consensus emerging that risk is the resultant of complex interactions among ecological and socioeconomic systems, and that the social construction of risk should be through the concept of vulnerability (IPCC 2012; Oppenheimer et al. 2014). Vulnerability and resilience increasingly require conceptualisation in terms of risks and we look at this issue in the South Asian context.

Rural-urban spaces are being re-defined in the South Asian context, where rapid urbanisation is projected and new types of land-use and economic activity are emerging in what were more traditionally defined spaces for such economic activity (Dasgupta et al. 2014). Similarly, the role of small scale technology in adaptation for alleviating climate change and the importance of the value of knowledge embedded in specific members of rural communities is increasingly recognised, for instance in the water sector such as rain water harvesting, irrigation, improved efficiency and water allocation rights (Hatcho et al. 2010; Rivera-Ferre et al. 2013; Ngoundo et al. 2007). Most of the recent revival in agricultural growth rates in India is being attributed to the growth in coarse cereals with renewed focus on dryland agriculture, rather than the contribution of intensive irrigation based dominant cereals such as rice and wheat. In this context, there is an increasing recognition of moving towards a risk minimization approach, with or without valuation, recognising the uncertainties of ecosystems, and in particular with regard to climate change and climate variability. India's drought prone areas programme, watershed development programmes are in the nature of welfare enhancement that is linked to recognition of trying to ensure thresholds levels of well being that minimise risk for the farmers concerned. Once we place risk minimisation at the centre of policy making, economic valuation can help in *identifying the damages or losses from risks and the costs of adopting strategies to reduce the risks.* Potentially large, irreparable losses or damages to the ecosystem pose severe risks, and may call for varied policy approaches primarily invoking the precautionary principle towards managing risks.

However, what is more widespread is the losses of ecosystem services which are often small, slow in their progress, and considered to be marginal and therefore tend to be neglected in economic decision-making. The impacts are ignored till these reach a magnitude which constitutes a severe risk or threat such as near complete loss of a water body such as a lake which has been dying gradually due to dumping of solid waste; or the accumulation of suspended particulate matter in a city from industrial pollution.

Economic valuation has been used extensively in *evaluating the impacts of such small perturbations,* those which are often slowly evolving over time and lead to gradual loss of ecosystem services. A range of valuation techniques have been used, focused primarily on monetizing values for both marketed and non-marketed or partially marketed ecosystem services. In the context of South Asia in particular, a large number of these studies have been motivated by the importance of natural resources in contributing to the livelihoods of communities dependent on them, and a recognition of the need to thereby conserve and nurture them, quite distinct from an international perspective such as the one on transboundary sharing of water resources or of the consequences of global warming. These studies span a range of sectors and find that values of ecosystem services are significant for sustaining livelihoods and should be accounted for in economic decision-making. A range of techniques have been used in valuing ecosystem services in the region—namely, Bangladesh, Pakistan, Nepal, India, and Sri Lanka (Dasgupta 2009; Haque et al. 2011); forests and common property resources (Chopra and Dasgupta 2008; Khan 2011), mangroves (Adhikari et al. 2010; Das 2011), wetlands and rivers (Alam 2008; Alam and Marinova 2003; Dasgupta et al. 2013b; Dehlavi and Adil 2011), floods (ADB 2011), pollution reduction (Adhikari 2012; Dasgupta 2004; Bogahawatte and Janaranjana 2011; Murty et al. 2011) and in specific contexts of distributional implications (Jacoby et al. 2011).

An international initiative, TEEB (The Economics of Ecosystems and Biodiversity) has put together the estimates based on different case studies in eleven different biomes (such as oceans, coral reefs, coastal wetlands etc.) for about 22 ecosystem services. It is clear that not all services are quantifiable and the estimates vary greatly across biomes. Figure 2.3 provides a sample from these estimates.

In some situations, risk management strategies offer win-win solutions and are pursued with zeal because these address a range of developmental goals, while concomitantly reducing stress on natural resources, such as reducing over harvesting of NTFPs or over grazing if alternative livelihood options open up. Examples of such management strategies include poverty alleviation or good governance. Examples of situations of synergies include the use of abatement technology that reduces local pollution and increases the availability of resources (clean air, safe drinking water) for immediate consumption, increasingly being termed as co-benefits in the literature.

There are also situations where the trade-offs are much sharper and economic valuation contributes by helping decision-makers internalize the environmental externalities. Here, informed and often tough choices are to be made such as the extent to which one should invest in climate resilient infrastructure given the uncertainties of the future. Economic valuation has been and continues to be an important contribution to decision-making.

	Open oceans	Coral reefs	Coastal systems	Coastal wetlands	Inland wetlands	Lakes and rivers	Tropical forests	Temperate and boreal forests	Woodlands	Grasslands
TOTAL	84	1195478	79580	215349	44597	13488	23222	4863	1950	3091
PROVISIONING SERVICES	22	20892	7549	8289	9709	5776	9384	1736	862	715
Food	22	3752	7517	2600	2090	196	1204	1204	203	82
Freshwater supply				4240	5189	5580	875	455		602
Raw materials		16792	32	1414	2430		3723	54	659	31
Genetic resources							1799			
Medicinal resources				35			1782	23		
Ornamental resources		348								
REGULATING SERVICES	62	33640	30451	135361	23018	4978	7135	456	1088	2067
Influence on air quality							957			
Climate regulation	55			4677	351		761	376	387	1661
Moderation of extreme events		33556		9729	4430		340			
Regulation of water flows					9369		36	3		
Waste treatment/water purification		77		120200	4280	4978	665	77	701	358
Erosion prevention				755			3211			47
Nutrient cycling and maintenance of soil fertility			30451		4588		1067			
Pollination							99			
Biological control	7	7								
HABITAT SERVICES	0	56137	164	68795	3471	0	5277	2575	0	298
Lifecycle maintenance			164	59645	917					
Gene pool protection		56137		9150	2554		5277	2575		298
CULTURAL SERVICES	0	1084809	41416	2904	8399	2733	1426	96	0	11
Aesthetic information		27317			3906					
Opportunities for recreation and tourism		1057492	41416	2904	3700	2733	1426	96		11
Inspiration for culture, art and design					793					

Fig. 2.3 TEEB estimates of the maximum value of 20 ecosystem services in 11 biomes (in 2007 dollars per hectare per year). *Note* Not all services were valued. *Source* Kumar (2010)

However and increasingly, the inadequacies of economic valuations in managing risks have led to demands for multi metric approaches, where some values are quantified, of which some are monetised, while a large quantum of values are qualitative in nature. These include the valuation of socio-cultural aspects of ecosystem services (Vihervaara et al. 2010; Martín-López et al. 2012; De Groot et al. 2010), and constitute a recognition of the fact that institutions within which values are articulated are themselves not ideologically neutral (Gómez-Baggethun et al. 2010). Hence, there is scope for better decision-making through complimentary processes of non-monetized valuation (Chan et al. 2012a, b). A recognition that is most important is that economic criteria can by themselves prove inadequate, and real time decision-making requires inputs from multiple disciplines. It is known that equity matters, with people showing distinct aversion to risk, and to inequality (Dietz and Atkinson 2010).

However, the challenge lies in articulating this in a meaningful manner in policies, that recognize that the socioeconomic and ecological systems are complex adaptive systems that are interdependent. For instance, it is important to integrate policy interventions that lead a social system to retain desired attributes or get back to a desired state after adverse perturbations. This would be a marker of increased resilience and robustness (Arrow et al. 2014).

2.4 Concluding Remarks: The Role of Ethical Considerations

We have argued in the preceding sections that ecosystems and economic systems are best perceived as complex adaptive systems, coevolving in different but interlinked ways. Two ways of designing policy with respect to these linkages are considered: the risk and management approach and the valuation approach. We argue that the two approaches can be used in different contexts and also complement each other in some. However, underlying both is an ethical concern with services and well-being in the future, both of humans and non-human species.

How we value the welfare of future generations and non-human species are the core questions which determine the stance we take as individuals, communities and governments. In economic decision making, in a limited manner, the social rate of discount reflects this value (Dasgupta 2008) and it is now accepted that lower rates of discount are to be applied to investments in natural resources which yield ecosystem services. In a manner of speaking a new norm has been created in the realm of knowledge for policy. Norms and values are created very slowly but they also stay ensconced for long. Some literature exists on how humans think beyond self interest in varying situations. Hodgson and Geoffrey (2012) for instance, holds that the interplay of self-interest and moral impulses has the potential to lead to changes in preferences and behaviour. When this happens, we will be able to move towards a rights-based approach to the sharing of some of our threatened resources and to a more stable co-evolution of economic systems and ecosystems. Meanwhile, we will need to depend on second best instruments such as risk analysis and management and valuation to move towards a balance in the relationships between ecosystems and economic systems as they co-evolve, moving, at times, in disparate trajectories.

References

ADB (2011) Pakistan floods: Preliminary damage and needs assessment, ADB. Available at: http://gfdrr.org/sites/gfdrr.org/files/Pakistan_Floods_2011_DNA_Report.pdf

ADB and IFPRI (2009) Building climate resilience in the agriculture sector in Asia and The Pacific. Asian Development Bank, Philippines, p 304

Adhikari N (2012) Measuring the health benefits from reducing air pollution in Kathmandu Valley. SANDEE working paper no. 69-12. SANDEE, Kathmandu

Adhikari B, Baig S, Iftikhar U (2010) The use and management of mangrove ecosystems in Pakistan. J Environ Dev 19 446–467

Alam K (2008) Cost-Benefit analysis of restoring Buriganga River, Bangladesh. Int J Water Resour Dev 24(4):593–607

Alam MK, Marinova D (2003) Measuring the total value of a river clean-up. Water Sci Technol 48(7):149–156

Arrow KJ, Ehrlich PR, Levin SA (2014) Some perspectives on linked ecosystems and socioeconomic systems. In: Barrett S, Maler K-G, Maskin E (eds) Environment and development economics: essays in honour of Sir Partha Dasgupta, OUP, UK, pp 95–116

Bogahawatte C, Janaranjana H (2011) Air quality and cement production: examining the implications of point source pollution in Sri Lanka. In: Haque AKE, Murty MN, Shyamsundar P (eds) Environmental valuation in South Asia. Cambridge University Press, Cambridge

Carpenter S (2001) Alternate states of ecosystems: evidence and its implication. In Press MC, Hantley N, Levin S (eds) Ecology, achievements and challenge. Basil Blackwell, London, pp 357–383

Chan KMA, Satterfield T, Goldstein J (2012a) Rethinking ecosystem services to better address and navigate cultural values. Ecol Econ 74:8–18

Chan KMA, Guerry AD, Balvanera P, Klain S, Satterfield T, Basurto X, Bostrom A et al (2012b) Where are cultural and social in ecosystem services? A framework for constructive engagement. Bioscience 62:744–756

Chopra K, Dasgupta P (2008) Assessing the economic and ecosystem services contribution of forests: issues in modelling, and an illustration. Int Forest Rev 10:376–386

Das S (2011) Can mangroves minimize property loss during big storms? An analysis of house damages due to the super cyclone in Orissa. In: Haque AKE, Murty N, Shyamsundar P (eds) Environmental valuation in South Asia. Cambridge University Press, Cambridge

Dasgupta P (2004) Valuing health damages from water pollution in Urban Delhi, India: a health production function approach. In: Environment and development economics, vol 9, no 1. Cambridge University Press, Cambridge

Dasgupta P (2008) Discounting climate change. J Risk Uncertainty 37:141–169

Dasgupta P (2009) Valuation of ecosystem services: methodologies, illustrations and use. In: Chopra K, Dayal V (eds) Handbook of environmental economics in India. Oxford University Press, Delhi

Dasgupta P, Kumari A, Bhattacharjee D (2013a) Socio-economic analysis of climate change impacts on foodgrain production in Indian states. Environ Dev 8:2013

Dasgupta P, Ghosh PK, Rahman R, Shah Alam Khan M (2013b) Mapping gains and losses from the Mayur ecosystem in Bangladesh. In: Narain V, Prakash A (eds) Water security in peri-urban South Asia, OUP, Dec 2015 (forthcoming)

Dasgupta P, Morton JM et al (2014) Rural areas. In: Field CB, Barrros VR et al (eds) Climate change 2014: Impacts, adaptation and vulnerability, report of the working group II, Intergorvernmental panel on climate change. Cambridge University Press, Cambridge

De Groot RS, Alkemade R, Braat L, Hein L, Willemen L (2010) Challenges in integrating the concept of ecosystem services and values in landscape planning, management and decision making. Ecol Complexity 7:260–272

Dehlavi A, Adil IH (2011) Valuing the recreational uses of Pakistan's wetlands: an application of the travel cost method. Sandee Working Paper no. 58-11, SANDEE, Kathmandu
Dietz S, Atkinson G (2010) The equity—efficiency trade-off in environmental policy: evidence from stated preferences. Land Econ 86:423–443
Expert Group (2009) Report of the expert group to review the methodology for estimation of poverty. Planning Commission, Government of India
Freeman AM (2003) The measurement of environmental and resource values: theory and methods. Resources for the Future, USA
Gale FP, M'Gonigle RM (2000) Nature, production, power: towards and ecological political economy. Edward Elgar Publishing UK
Gómez-Baggethun EG, Mingorría S, Reyes-García V, Calvet L, Montes C (2010) Traditional ecological knowledge trends in the transition to a market economy: empirical study in the Doñana natural areas. Conserv Biol 24:721–729
Haque AKE, Murty MN, Shyamsundar P (eds) (2011) Environmental valuation in South Asia. Cambridge University Press, Cambridge
Hatcho N, Ochi S, Matsuno Y (2010) The evolution of irrigation development in Monsoon Asia and historical lessons. Irrig Drainage 59:4–16
Hodgson, Geoffrey M (2012) From pleasure machines to moral communities: an evolutionary economics without homo economicus. University of Chicago Press, Chicago
IPCC (2012) Managing the risks of extreme events and disasters to advance climate change adaptation. In: Field CB, Barros VR, Dokken DJ et al. (eds) A special report of the working groups I and II of the Intergovernmental panel on climate change. Cambridge University Press, Cambridge, United Kingdom and New York, NY, USA, 582
IPCC (2014) Summary for policymakers. In: Field CB, Barros VR, Dokken DJ et al. (eds) Climate Change 2014: Impacts, Adaptation, and Vulnerability. Part A: Global and Sectoral Aspects. Contribution of Working Group II to the Fifth Assessment Report of the Intergovernmental Panel on Climate Change Cambridge University Press, Cambridge, United Kingdom and New York, NY, USA, 1–32
Khan H (2011) Valuation of recreational amenities from environmental resources: the case of two national parks in Northern Pakistan. In: Haque AKE, Murty MN, Shyamsundar P (eds) Environmental valuation in South Asia. Cambridge University Press, Cambridge
Kumar KS, Kavi, Parikh J (2001) Indian agriculture and climate sensitivity. Global Environmental Change 11(2):147–154
Jacoby H, Rabassa M, Skoufias E (2011) Distributional implications of climate change in India. Policy Research Working Paper, 5623, Poverty Reduction and Economic Management Network, The World Bank
Kumar P (ed) (2010) The economics of ecosystems and biodiversity. Earthscan, London
Lele S, Srinivasan V (2013) Disaggregated economic impact analysis incorporating ecological and social trade-offs and techno-institutional context: a case from the Western Ghats of India. Ecol Econ 91:98–112
Maler KG (1999) Deep economics of shallow lakes. Economic-ecological modelling. Mimeo Lecture delivered at Institute of Economic Growth, Delhi
Marshall A (1890) Principles of economics. McMillan and Co., New York
Martinez-Alier J (1991) Environmental policy and distributional conflicts, In Robert Costanza (ed.), Ecological Economics: The Science and Management of Sustainability. Columbia University Press 118–136
Martinez-Alier J et al (2010) Social metabolism, ecological distribution conflicts, and valuation languages. Ecol Econ 70:153–158
Martín-López B et al (2012) Uncovering ecosystem service bundles through social preferences. PLoS one, 7(6)
Millennium Ecosystem Assessment (2005) Ecosystems and human well-being: synthesis. Island Press, Washington, DC
Mukhopadhyaya P, Kadekodi GK (2012) Missing the wood for the ore. Econ Polit Wkly 46:61–67

Murty MN, Gulati SC, Banerjee A (2011) Benefits from reduced air pollution in Delhi and Kolkata: a hedonic property price approach. In: Haque AKE, Murty MN, Shyamsundar P (eds) Environmental valuation in South Asia. Cambridge University Press, Cambridge
Ngoundo M, Kan CE, Chang Y, Tsai SL, Tsou I (2007) Options for water saving in tropical humid and semi-arid regions using optimum compost application rates. Irrig Drainage 56:87–88
Oppenheimer M, Campos M, Warren R, Birkman J et al (2014) Emergent risks and key vulnerabilities. In: Climate change 2014: impacts, adaptation and vulnerability. Part A: global and sectoral aspects, contribution of WGII to the fifth assessment report of the IPCC. Cambridge University Press, UK, pp 1039–1099
Pagiola S, von Ritter K, Bishop J (2004) Assessing the economic value of ecosystem conservation environment-development paper no. 101, the World Bank in collaboration with Nature Conservancy and IUCN, the WCU, Washington DC
Puroshothaman S et al (2013) Beyond money metrics: alternative approaches to conceptualising and assessing ecosystem services. Conservation and Society 11(4):321–325
Rivera-Ferre MG, Di Masso M, Mailhost M, Lopez-i-Gelats F et al (2013) Understanding the role of local and traditional agricultural knowledge in a changing world climate: the case of the Indo-Gangetic Plains. In: CGIAR-CCAFS Program, Nepal, p 98
Rockstorm J et al (2009) A safe operating space for humanity. Nature 461:472–475
Scheffer M (1997) The ecology of shallow lakes. Mcgraw Hill, New York
Smith A (1776) An enquiry into the nature and causes of the wealth of Nations. Bibliobazaar 2008
Stern N (2007) The economics of climate change: the stern review. Cambridge University Press, Cambridge, UK
Vatn A (2009) An institutional analysis of methods for environmental appraisal. Ecol Econ 68:2207–2215
Vihervaara P, Rönkä M, Walls M (2010) Trends in ecosystem service research: early steps and current drivers. Ambio 39:314–324
Walras L (1877) Elements of theoretical elements, or the theory of social wealth (2014) (trans: Donald A Walker and Jan van Daal). Cambridge University Press, Cambridge
Wilson EO (2001) The diversity of life. Penguin Books, London

Part II
Ecosystems' Perspectives

Chapter 3
Interactions Between Humans and Ecosystems in Himalayas of India and Its Socioeconomic and Ecological Consequences: An Ecological Modelling Approach

Sunil Nautiyal

3.1 Introduction

The ecological models are extensively accepted because they are tools for managing the future ecosystem/landscape and they provide enough opportunity to study multifunctional approaches such as price, policy, trade, etc. of the system. The effectiveness of the natural resource management requires a detailed understanding of the patterns and processes that exist within both the natural system itself and the human institutions associated with the utilization of the resource (Deadman 1999). The applicability of the model related to the natural resource management application should ideally incorporate the human system associated with that system. This approach provides enough opportunity to interact and observe the relationship between interactions of behavior of complex humans and ecosystems in any region and might be helpful to simulate the change in human behavior in due course and patterns and implications of landscape change. Natural landscapes are an important ecological, economic, and social/cultural resource that gives the basis for the sustainability of any region and contributes significantly to the quality of life of the local people. The degree of naturalness in the landscape has, in general, a positive effect on the quality of the landscape (Waikato 2003; Brabyn 2005).

Research in natural resource management may be characterized as a search for an understanding of patterns and processes relating to a particular resource

S. Nautiyal (✉)
Centre for Ecological Economics and Natural Resources (CEENR),
Institute for Social and Economic Change (ISEC), Bangalore, India
e-mail: nautiyal_sunil@yahoo.com

H. Sandhu (ed.), *Ecosystem Functions and Management*,
DOI 10.1007/978-3-319-53967-6_3

(Anderson and Evans 1994) and this relies on the use of resource, economics, and a production system of the landscape which, in most of the cases, is available in many non-comparable scales (Mallawaarachchi et al. 1996). This is the main hurdle for using data for detailed modeling and analysis. But in our study, we have gone in detail to analyze the interaction between humans and ecosystems in a part of the Himalayas of India. In our approach and methodology, we tried to articulate the human decision-making behavior particularly to secure his livelihood in the fragile Himalayan environment and tried to model the whole scenario development to analyze the patterns of landscape change. The work on the development of a model framework and modeling for natural resource management in the Himalayan region is still very limited. Therefore, the present study aims to understand the development process for the model for natural resource management for the Himalayas in India based on the *AGENT's behavior under changing socio-economic and environmental conditions. In the region, there are six types of decision alternatives which the agent has been selecting deliberately for last three or four decades to secure livelihood. The decision alternative was defined precisely by Schleiffer (2005). The aim of this research was to design and develop a new modeling approach for the complex system analysis in the Himalayan landscape. The objectives for the study are (A) to study the human-ecosystem interaction and development of the framework of integrated natural resource management model using the existing data sets, and (B) testing the efficiency of some known natural resource management models for their applicability in mountain regions.

3.2 Study Area

The study area is the central part of the Indian Himalayan region located between 30017′N-30041′N latitude and 79040′E-8005′E longitude. The whole region is divided in three agro-ecological zones: Lower elevation: >1000 masl; Middle elevation: between 1000 and 1800 masl. The current study was executed in the higher elevational zone of the Central Himalayas and represented the whole higher elevational zone in which the area is located. The area has a great ecological importance in terms of diversity of the natural resources. The world famous national park, “Valley of Flowers,” and a Himalayan Biosphere Reserve, “Nanda Devi,” were established in this region for the conservation and management of biodiversity and ecosystem services and cover 223,674 ha landscape of the region. The overview of the area is given in Fig. 3.1. There are seventeen villages' in the study area, of which four are totally uninhabited. The total population of the villages was 2762 (947 male adults, 781 female adults, and 877 children below 15 years of age).

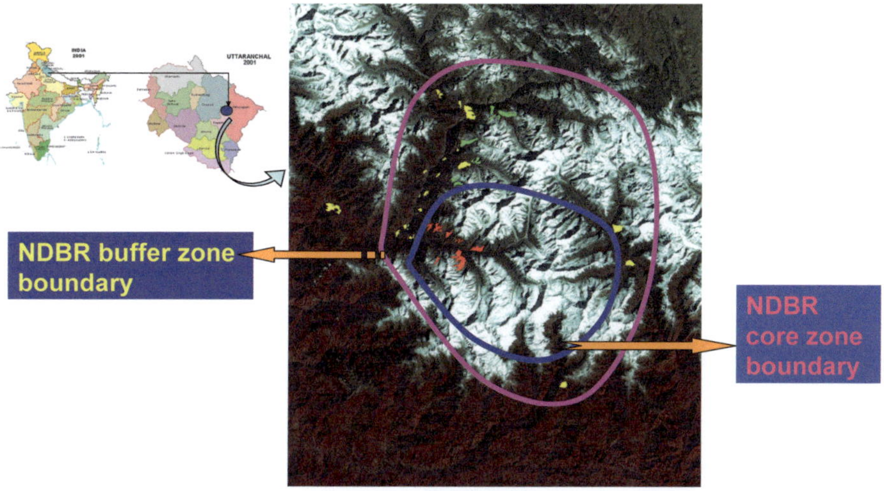

Fig. 3.1 Landsat-7 ETM+ image of the area also covering a Himalayan biosphere Nanda Devi (*yellow* polygons are villages; *red* polygons-pasture included banned for grazing; and *green* polygons-pasture available for grazing)

3.3 General Description of Ecosystems in the Study Area Located in the Indian Himalayan Region

The ecosystems in the study area are similar to that of ecosystems in the Indian Himalayas where agriculture, animal husbandry, forests, and domestic sub-systems are interlinked and have a flow of resources with the market. To analyze and understand the land-use, land-cover change and developmental scenario in the region, data pertaining to agriculture, including medicinal plants cultivation, animal husbandry, natural resource utilization patterns, forest and domestic sectors, and socio-economic profiles of local people are analysed at three points in time (i.e., 1970–80, 1980–90, and 1990–2000). The study is important because, based on the ecological and economic analysis of the rural ecosystem, land-use information is produced to understand the causes and consequences of scenario changes and developmental processes in the region.

3.4 Methodology

To understand the process of how the people have been changing their lives and activities due to a variety of factors (such as environmental, policy, socio-economic), and therefore, how the natural ecosystems also change with human activities need to have a long-term study of any region. The work done for last 11 years in the Himalayan region of India facilitates the start of work in this

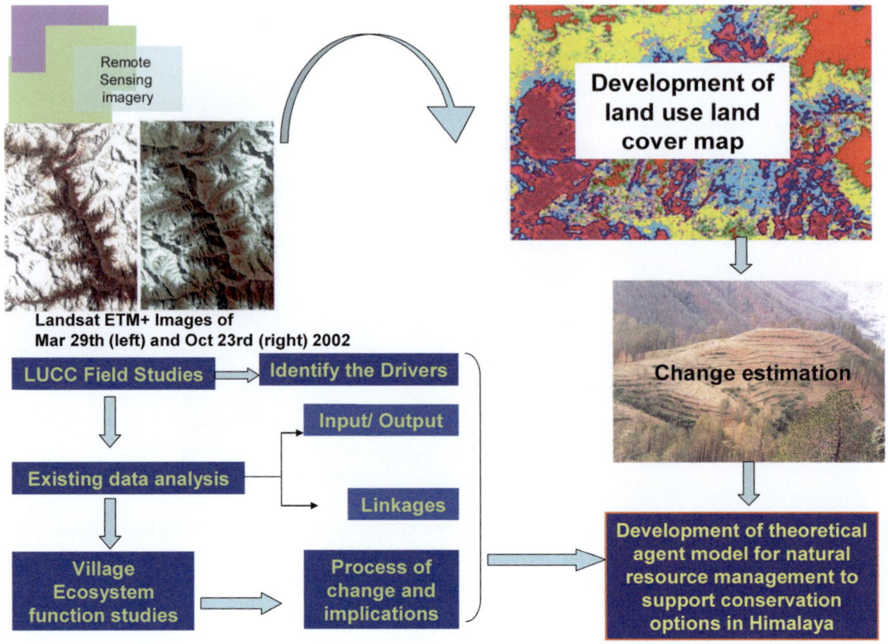

Fig. 3.2 Work-flow of methodology for the study (Nautiyal and Kaechele 2009)

direction, and thus the work-plan for the data analysis was developed (Fig. 3.2). The work plan was categorised into two segments:

- Analysis of existing data sets.
- Analysis of remote sensing imagery.

3.4.1 Analysis of Existing Data Sets

This is an interdisciplinary study to bring together ecological sciences with social sciences. The data analysis was done in a detailed way for the whole region over three points of time, and again, we formed the study in a microscopic view point to understand the basic idea of how the farmer made his or her own decisions in the system. The analysis approach was particularly exemplary because it has tried to analyze different food production systems of the landscape such as agriculture, medicinal plants cultivation, kitchen gardens, and animal husbandry using the currencies in ecological, economical, and social terms of scenario development. Further, it also analyzes the inter-linkages and interdependencies of all the sectors of the landscape. This is an approaching effort to strengthen conservation programs for biodiversity on one hand and development of the local economy on the other in the fragile Himalayan environment. The methodology pertaining to the study of agriculture, animal husbandry, a natural resource utilization pattern, and the whole village ecosystem function is described in detail in Nautiyal 1999; Nautiyal et al. 2002, 2003a, b; and Rao et al. 2005.

3.4.2 Analysis of Remote Sensing Imagery

To understand that how the sustainable ecosystems (including the interaction between anthropogenic and natural resource systems) are altered from past decades, we need to emphasize a holistic or integrating approach. In this endeavor, the analysis of remote sensing imagery gives geographic and temporal information for land-use and land-cover change analysis and is helpful in understanding the degree of extent of human dimensions to environmental change and the outcome of human actions across the landscape. The imagery for the current study is used two times a year, particularly during a two-crop calendar (winter and summer) of the region in a decade interval starting in 1972. The acquisition dates of the imagery are [11-02-1973; 26-10-1972; 20-04-1979; 26-10-1979 (Multi-Spectral Scanner-MSS); 21-05-1992; 22-10-1992 (Linear Imaging and Self Scanning-LISS-1); 29-03-2002; 23-10-2002 (Enhanced Thematic Mapper-ETM+)]. The image calibration was done and all the images were registered manually using ENVI image to map registration method. A satellite differential GPS was used in the field (accuracy $\approx$ 2 m) to collect geo-referenced information for geometric correction of the satellite imagery using the ground control points (GCPs). Easily recognizable landscape features such as confluence of the rivers, water channels, agriculture, village, different forest types, etc., were recorded and used to correct the image with the help of ArcGIS 9.0. The base image was calibrated with the landscape of the region while using GPS data of the area (accuracy $\approx$ 2 m) to standardized the satellite data. For this approach, we have selected the ETM+ imagery of the region as a base image of the area. Thereafter all the imagery registered manually with the ENVI 4.1. To test the efficiency of the natural resource management model, we analyzed the primary data linked with the satellite data of the region. In this approach, we have selected the animal husbandry linking with the alpine pasture ecosystem of the region. We used Landsat-3 MSS (path/row, 156/039) and Landsat-7 ETM+ satellite image (path/row, 145/039) to cover the study area. All the information generated here is interpreted with GCPs, field data, field observation, and in-depth knowledge of the area and plant communities.

Development of a model framework for natural resource management in the Indian Himalayan region was done once the methodology was synthesized. Then a plan was designed for the arrangement of the data for the study. The work-flow for the methodology is given in Fig. 3.2.

3.5 Results and Discussion

3.5.1 Development of Model Framework

The entire regional system is modeled after a collection of decision-making entities by the local farmer, called agents. The local farmer individually assesses and monitors his own situation and makes decisions based on a set of rules. The driving forces are the key factors which influence the activities of the agent to behave in time, to time. The agent may execute various behaviors according to the system, for example, changes in the production system in the form of land intensification, resource collection, consumption, and selling the produces. We have found the approaches simpler for the mountainous region, as this model consists of behaviors of the decision maker and relationships between different production systems. The framework of model is simpler, however, and can exhibit the complexity of the system. The framework was developed based on the information produced for the real world and the decision-making process of the decision maker based on information available to him, from both the points in time and space. An overview of the framework of the agent model for natural resource management is depicted in Fig. 3.3.

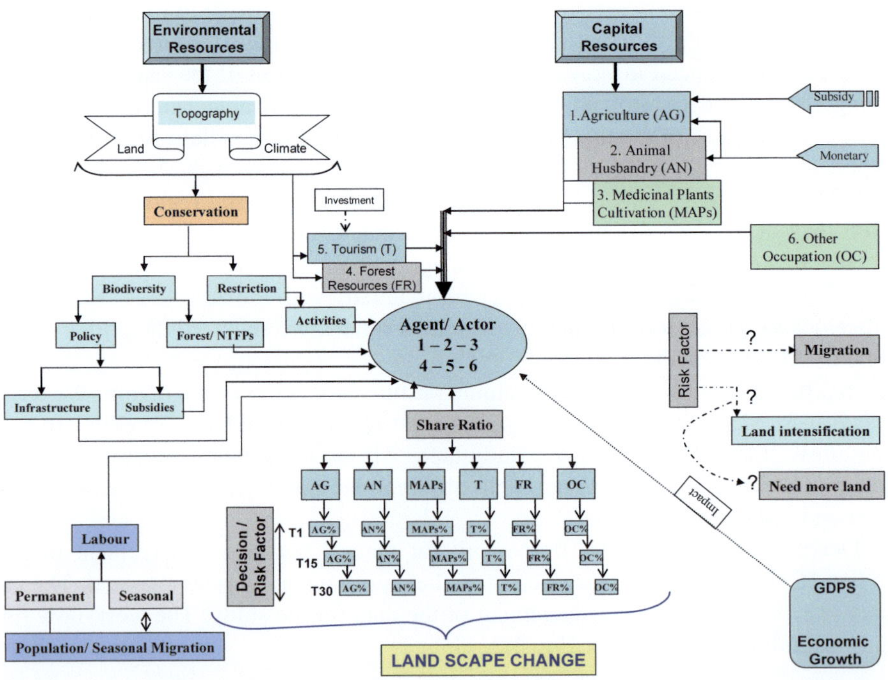

Fig. 3.3 Overview of the framework of the AGENT model for natural resource management (Nautiyal 2011a)

3.5.2 Decision Making Behavior in Uncertainty

The data are analyzed to understand that how human behavior is changing under changing socio-economic and environmental conditions. The agent is trying to explore the sources of income to sustain his life in the fragile environmental condition. For the current analysis, the period of 1970–80 is the starting period to study agent behavior and ecosystem functioning of the region. Therefore, in the context of time (t1), it is important to analyze the system's functioning and consequently the behavior of the agent of the area. There are four main activities of the people where a local economy is centered. The temporal data show the activities of people in the region (Fig. 3.4). The demographic study at very beginning was helpful to design the work plan for further study. The working population for an agriculture sector is straight for all the periods. This is because every household has agriculture land and women, particularly, contribute to this sector throughout the year. Due to strict policies, tourism is no longer in the region. However, the involvement in animal husbandry is declining and interest in other occupations (such as daily wage labor and service) is increasing. Household behavior and projections are gaining more importance. This holds not only for the projections, but also the decision-making behavior for the scenario development. Therefore, there is an increasing interest in micro-oriented interdisciplinary research, which proves that household processes have a large influence on economical and ecological processes (Nelissen 1991; Klevmarken 1983; Rao et al. 2005; Bisht et al. 2006).

3.5.3 Land-Use Intensification and Productivity

In the region, the rainfed agriculture on steep terraces is the predominant form of land use. Most the population is involved in agriculture. Average per capita land

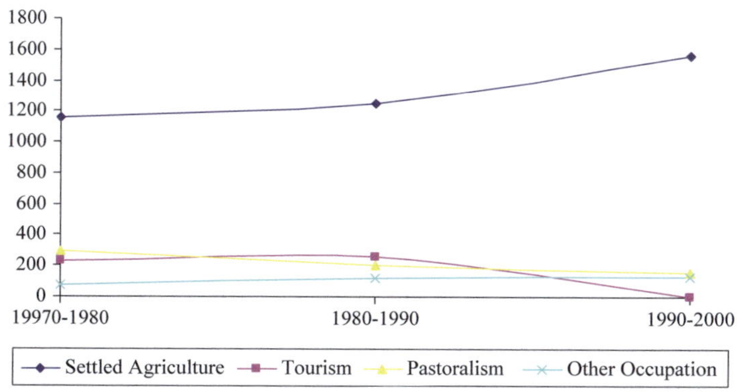

Fig. 3.4 Activities of people at different points of time in the region

holding ranges from 0.15 to 0.19 ha. The magnitude of land-use change in the region comes from two types: (a) where traditional land use is totally changed and farmers have introduced modern crops such as tomato and bell pepper cultivation to fetch more economic benefit; (b) where traditional land race-based cultivation is being replaced by the high yielding varieties of paddy and wheat and other cash crops. But the former is restricted to valleys of low and middle elevational zone and so far not adopted by the people of high elevational zone. The land use change trend was analyzed from 1970 to present. It is very difficult to analyze the land-use change pattern before 1970 due to some flaw in the methodology, and more particularly, only using the conventional methods. Therefore, we assumed that during 1970 the land-use was optimum under each crop of the area and that it started changing thereafter. Based on the study, we found that the traditional land-use system is changing very fast in the region and intensifying because the introduced crops/high yielding crop varieties fetch more economic benefit. From the period of 1970–80 and 1980–90, the land-use under many traditional crops was reduced between 15 and 60%. However, this trend continued, and in the year 2000, the decline in land-use under traditional crops increased between 50 and 96% (data not presented). Meanwhile, the productivity per capita per year from a hectare of agricultural land has increased 286 kg (1970–80) to 394 kg (1980–90) to 579 kg (1990–2000) (Fig. 3.5).

In the Himalayan region, the agriculture production is directly linked to surrounding ecosystems; therefore, equilibrium should be maintained between agricultural production systems and surrounding natural ecosystems for evaluation of sustainability (Lefroy et al. 2000; von Wiren-Lehr 2001). An extension of agricultural land-use into forestland, coupled with a variety of factors such as fragile land-use policies, lack of awareness among the people, ecosystem degradation arising from traditional practices of litter collection, for maintaining agricultural soil fertility and unsustainable harvesting and overexploitation of the resources could be a threat to forest biodiversity and ecosystem services on one hand and sustainability of traditional farming on the other (Singh et al. 1984; Pilbeam et al. 2000).

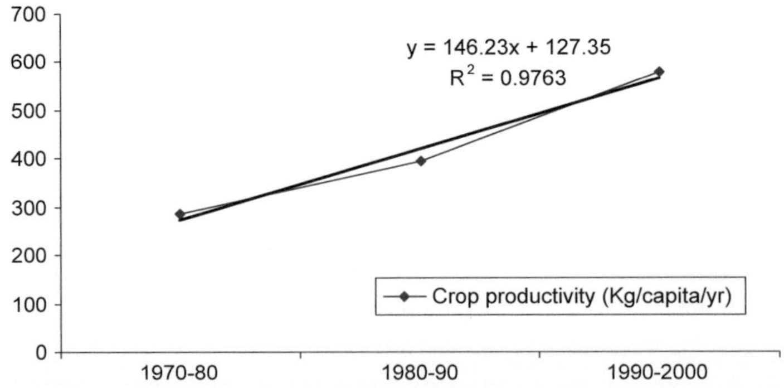

Fig. 3.5 The per capita per year productivity from agricultural land at different points of time

3.5.4 Resource Collection and Dependency

The scenario development is showing an indication of a threat to sustainable land-use development in the region. To obtain good monetary output from the agriculture sector, the farmer is changing the land-use and promoting the crops having good market value. Due to this the fodder (crop by product), yield has been decreased up to 35%, and in view of that, the pressure on forests for fodder collection and leaf litter for manure preparation is increased. A similar trend was proved in the study done by Semwal et al. (2004) in a watershed of the Central Himalayas. The perseverance of a farmer to secure the optimum output per capita per year has led to a collection of leaf litter from the forest increasing 222 kg/year (t3) from 100 kg/year (t1). The different forest resource collection from the forest at different points of time is depicted in Fig. 3.6.

3.5.5 Options to Secure the Livelihood in Mountains

In the four major activities (agriculture, animal husbandry, forest resource collection and tourism), there are six options such as agriculture (AG), animal husbandry (AH), medicinal plants cultivation (MAPs), non-timber forest products collection (NTFPs), tourism (T), and other occupations (Ot) such as daily wage labour, service, etc., for the agent to sustain the livelihood in the region (Fig. 3.7). The system analysis showed that at starting point (t1), the maximum share in income among the sources of the agent or decision maker was coming from animal husbandry (40%) followed by tourism (35.2%), agriculture (14%), NTFPs collection (10%), other occupation (0.6%) and the least was from medicinal and aromatic plants cultivation (0.2%). However, due to a variety of factors, such as implementation of NRMm (conservation policies), a ban on NTFPs collection, a ban of tourism, restriction of grazing, socio-economic and cultural changes, migration, and overall scenario

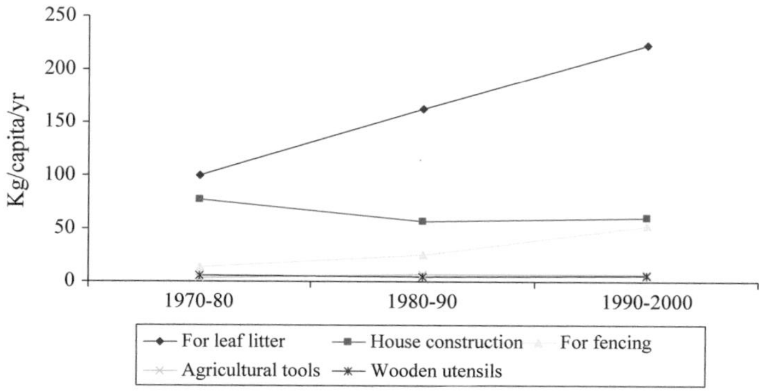

Fig. 3.6 The resource collection from the forests for agriculture and domestic needs

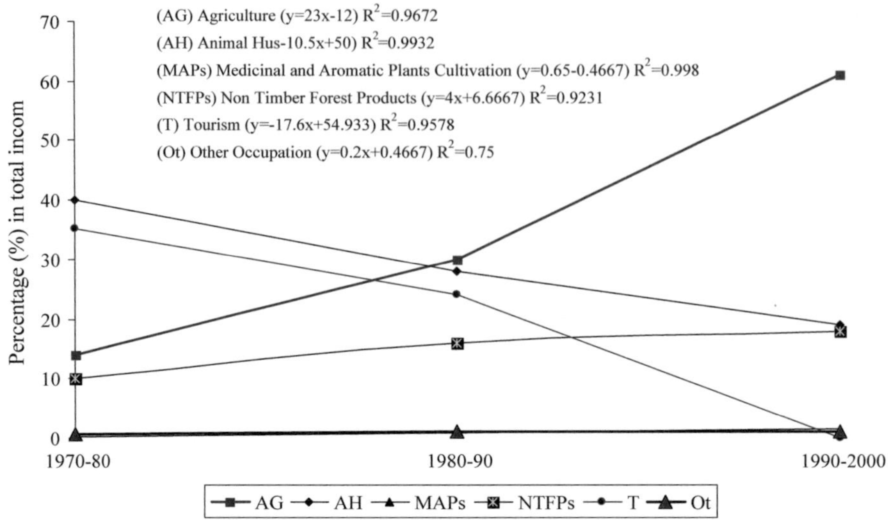

Fig. 3.7 Share of different sector in household income for last three decades at intervals of a decade

development, etc., at the point of t3 the agriculture sector is contributing more than half (61%) in the total income coming from the different sources, followed by animal husbandry (19%), forest resources (18%) and the least from other occupations (1%).

Meanwhile, to replenish the loss from the tourism and animal husbandry, the agent or decision maker is trying to acclimate himself in the changing environmental condition. Besides, with an emphasis on increasing the productivity of agricultural crops, his interest towards medicinal and aromatic plants cultivation is also increasing and consequently, he increased the share ratio from 0.2 to 1.3% in the medicinal and aromatic plants cultivation sector.

These are low volume and high value crops, but the current scenario holds the hurdle to expand the area under cultivation of medicinal and aromatic plants (Maikhuri et al. 1998; Nautiyal et al. 1998). Due to a variety of factors such as policies, exploitation by middlemen traders, and banning imports on high value crops, the decision maker is encouraged to cultivate cash crops in large land area (Midmore et al. 1996; Semwal et al. 2004) and making hurdle to expand area of low volume high value such as medicinal and aromatic plants cultivation in a mountainous region. The increasing area of cultivation of potatoes, amaranth, kidney beans, etc., in the region is driven by the socio-cultural change from subsistence to market economy facilitated by improvements in accessibility and supplies of staple food grains to the local market (Maikhuri et al. 1996; Semwal et al. 2004).

3.5.6 Change in Land-Use Land Cover

The overall system analysis concluded that the study of spatial and temporal distribution of land-use and land-cover change is important. This is important to understand the change in human behaviour due to variety of factors and consequences to this how land landscape is changing. For landscape change analysis data from remote sensing is helpful to monitor the current state and changes to the land-use land-cover in the fragile Himalayan environment. Using the satellite data available, we have developed the land-use land-cover map and studied the land-use cover change detection in a part of Himalaya. In this endeavor, the study was done in the two-cropping season starting in 1972–2002, at a decade interval. For the present study, the land-use cover map of the area for the period of October 1979 and October 2002 is depicted in Fig. 3.8a, b. The classification was done while following the supervised and decision tree classification methods. Based on satellite data interpretation and visual observation, we found that most of the agricultural expansion occurred in forest/open land adjacent to agricultural land, followed by reserve forests. The forests' cover shows decline in the area of different forest types (Fig. 3.9). The trend of expansion in agricultural lands and narrowing the forest cover is a common trend in the Himalayas of India (Virgo and Subba 1994; Thapa and Weber 1995; Sen et al. 2002; Semwal et al. 2004) and other mountainous regions in developing countries (Fox et al. 1995; Midmore et al. 1996; Jianchu et al. 2005; Thongmanivong et al. 2005). Concentration of agricultural expansion in higher altitudes of a watershed of Central Himalaya, where ecological conditions are more favorable for the most profitable cash crops (Sen et al. 2002), indicates that expansion is driven largely by farmers' tendency for maximization of income and expansion occurred largely in forestland on moderate slopes where traditional terracing was feasible and institutional arrangements of forest protection provided in forest policy were weak (Semwal et al. 2004).

Based on the current analysis, the sustainable land-use development to support the livelihood of the local farmer and conservation options in the Himalayan region should be given prime importance. Sustainable land-use is crucial not only for the sustainable livelihood of the Himalayan/mountain people, but also the numerous other people living in the adjoining Indo-Gangetic plains, as accelerated erosion due to inappropriate land-use in the Himalayas partly contributes to devastating floods in the plains (Ives and Messerli 1989; Saxena et al. 2001; Semwal et al. 2004). Efforts have been made to analyze changes in broad land-use/land-cover types in the Himalayas (Virgo and Subba 1994; Thapa and Weber 1995; Schweik et al. 1997; Jackson et al. 1998; Rao and Pant 2001; Gautam et al. 2002). Yet, knowledge on changes in spatial patterns of agricultural land-use, driving factors, and their implications within the context of sustainable development is limited (Thapa 1996; Hurni 2000; Sankhayan et al. 2003). Therefore, the in-depth study for sustainable land-use development will be very helpful in designing and developing the strategies for sustainable development of Himalayan agriculture and natural resources (Nautiyal et al. 2005).

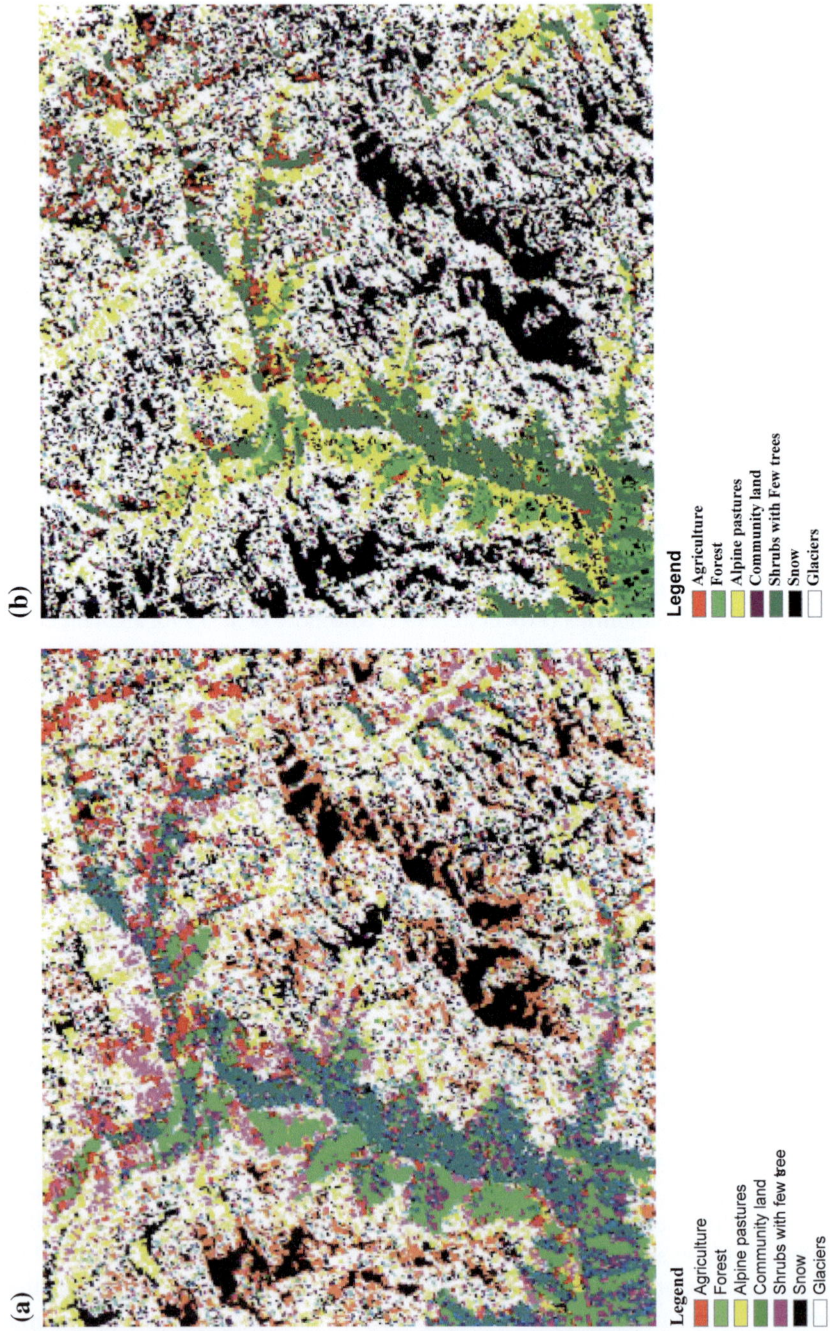

Fig. 3.8 **a**, **b** show the land-use land-cover map of the study area for the period of 1979 and 2002, respectively

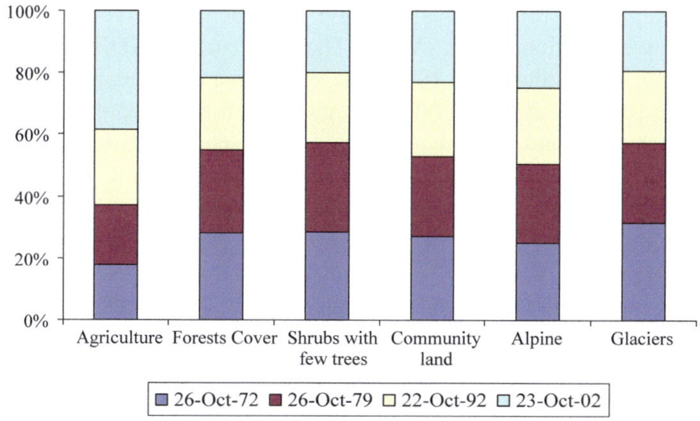

Fig. 3.9 Percent distribution of land use land cover at different points of time in study region of Himalaya

3.5.7 Testing the Efficiency of Some Known Natural Resource Management Models (NRMm) for Their Applicability in Mountain Regions

This is an important objective of the study, because we have verified the existing natural resource management model (NRMm) in the region. Over the last few decades, there have been several models for conservation and management of the resources. The network of protected areas such as the biosphere reserve or national parks are the example of the few NRMm where the conservation policies have been executed for maintaining the ecosystem's health, but questioned with a variety of problems, and the most widespread one commonly termed as resource-people conflicts. We have verified the model in an economic and ecological point of view to see the impact of grazing on alpine pasture while arguing whether grazing is good or bad for a pasture ecosystem. In the rural ecosystem, the animal husbandry sector provides subsidiary income to local people after the agriculture land use. But unfortunately, the animal husbandry sector is a badly affected sector due to the implementation of a natural resource management model, however unintentionally. The animal husbandry system, which was strong during 1970–75 and contributed to the sustenance of the local people, started declining after the implementation of conservation policies. Therefore, the economic consequences of the implementation of NRMm are showing a negative trend in contributing to the economic support of the livelihood of the local farmer from this sector. Meanwhile, it is important to analyze the ecological perspective of the implementation of NRMm. In this endeavor, a study based on satellite data is done on the extraction of vegetation cover in different alpine pastures of Central Himalaya to verify the efficiency of the natural resource management model for the analysis of the ecosystem's health. This study was undertaken to view the impact livestock grazing in a core zone and buffer zone of the alpine pastures of Nanda Devi Biosphere Reserve (NDBR) and the Valley of Flowers National Park (VOF). The core zone of NDBR and VOF were strictly banned for grazing since 1982. The vegetation cover maps were developed for the alpine pastures of whole NDBR region when there were no boundaries of core and buffer zones and for core and buffer zones separately after implementation of NRMm in 1979 and 2002. The importance of this study is the subject of whether the livestock grazing should be open or not in protected areas (core zone of protected areas) of the region. However, the vegetation map for 2002 developed by us for the pastures of buffer and core zones of the reserves shows that the habitat of bushes/thorny bushes is getting more expansion in their establishment in the core zone of NDBR and VOF in comparison to buffer zones of NDBR, area before implementation of NRM in the region (data not presented*).

We have concluded that in economical and ecological view points, the existing NRMm are not suitable for the region. Many researchers working in the region stated that the diversity of any area is depends on the degree of protection given to it. If an area is protected for the name of conservation, then there are always the chances for low density and diversity of the species. The low species richness and low density is the result of the high regeneration potential and well-developed canopy of several dominant species after giving strict protection to any natural environment (Negi and Nautiyal 2005). For the past two decades, though, there have been several reports about weeds choking the plant diversity of the VOF, and researchers blame the weed *Polygonum polystachyum* (Kala 2002). Remote sensing techniques, including visual observation and image interpretation, are useful for providing valuable information in a short amount of time for large areas (Sandstrom et al. 2003). Many studies indicated that livestock grazing in alpine pastures maintain the diversity of the ecosystems/landscape (Singh and Kaur 1983; Naithani et al. 1992). However, policy makers think that the pastoral practices are responsible for ecosystem degradation (Chakravarty-Kaul 1998). Therefore, the scientific studies will be very helpful in understanding the complexity of the system (Kala 2004) and designing and developing the effective conservation program for the conservation and management of alpine ecosystem. The Hjortso et al. (2006) have pointed out that the conservation and management would be effective and feasible if system function and structure related to the restriction and non-restriction are perceived in a comprehensive holistic way, integrating social and biological relations and dependencies.

3.6 Conclusion

Through the in-depth field investigation and remote sensing image analysis, the paper indicated that the sustainable resource development and utilisation of any area depends on the socioeconomic condition of the inhabitants of that region. In order to evaluate the sustainable resource management and economic development, a novel agent model has been designed based on the complex interaction between human and ecosystem in a part of Himalayas of India. In this system approach there is a lot of opportunity for the researchers/scientists from different disciplines to come on one ground for the interdisciplinary work. This is also important to maintain and store the information from different aspect in one place and reduce the risk of segregation/fragmentation. This modeling approach will have the main advantages in terms of (i): to simplify the complex human ecosystem interactions while analyzing and evaluating the basic approaches of complexity; (ii): flexibility in selection of individual objective for the systems analysis and making the avenues/platform to link other objectives of the system from very simple to very complex, depending on availability of data and approach of the study; (iii): helpful in making the bridge among the research groups, policy makers and local people at different phase of the development.

For the last several decades, many programs have been implemented for the conservation of biodiversity while giving less importance for the reciprocal relationship between humans and adjacent ecosystems. The conservation and management of the ecosystems/landscape and development of the local economy are interwoven. The complexity of the system is a major hurdle in achieving the goal —"conservation and compensation." However, the long-term study of the representative sites of the rural ecosystems and landscape is key factor in understanding the system functioning and designing the appropriate strategies for sustainable development of the region. Conservation and economic security are important at the same spatial and temporal dimension.

The following points are suggested based on this study:

1. rather than a strict ban for conservation of the resources, the appropriate design and framework should be developed for sustainable utilization of the resources. This should be based on the scientific study undertaken in any area. The increment in resource collection to sustain the livelihood of the local people and the expansion of the other species over the native herbaceous vegetation in the alpine pasture is an example of the "conservation" approach of most of the developmental models in the Himalayan region. However, the sustainable management of the resources and development would be possible at same time if the way chalk out for better economic development from the better ecological performance (Xu et al. 2006).
2. There are always gaps between the three-dimensional approaches of the sustainable development (such as ecological, economical, and social perspective of the region). Therefore, designing and development of the model framework and the interdisciplinary study, while inviting experts from every field for developing the appropriate strategies, is equally important (Fig. 3.10).
3. If income of the local people is secure while providing incentives to use the resources such as grazing and other non-vegetation related activities under the parameter of environmental conservation, then the chances of conservation and management of the ecosystems is possible.

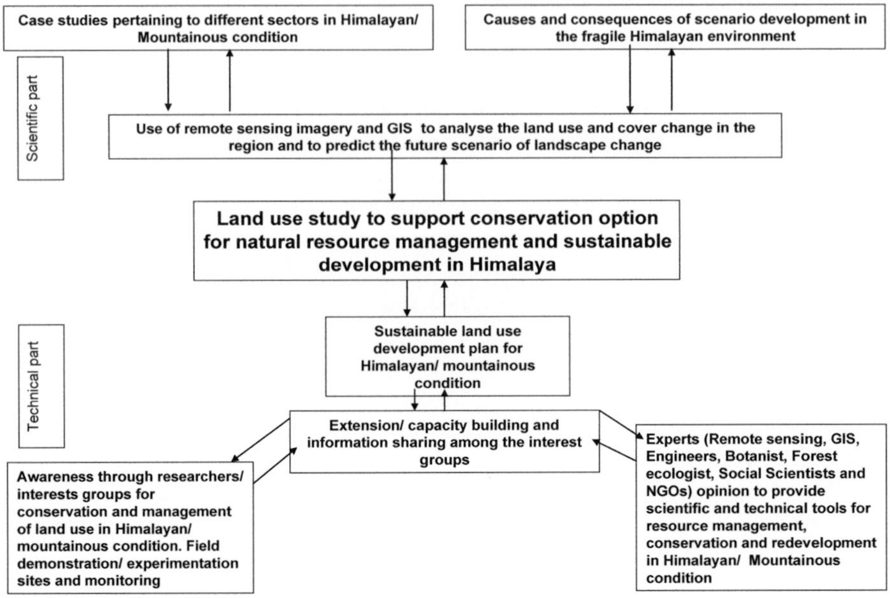

Fig. 3.10 Sustainable land-use development strategies for natural resource management in the Indian Himalayan region (Nautiyal 2011b)

4. The probability for unsustainable land-use and overexploitation of the resources maximizes several times, which leads to severe environmental degradation in the Himalayan region.

References

Anderson J, Evans M (1994) Intelligent agent modeling for natural resource management. Math Comput Model 20:109–119

Bisht IS, Rao KS, Bhandari DC, Nautiyal S, Maikhuri RK, Dhillon BS (2006) A suitable site for in-situ (on-farm) management of plant diversity in traditional agroecosystems of western Himalaya in Uttaranchal state: a case study. In: Genetic resource and crop evolution

Brabyn L (2005) Solutions for characterising natural landscapes in New Zealand using geographical information systems. J Environ Manag 76:23–34

Chakravarty-Kaul M (1998) Transhumance and customary pastoral rights in Himachal Pradesh: claiming for high pastures for Gaddis. Mt Res Dev 18:5–17

Deadman PJ (1999) Modeling individual behaviour and group performance in an intelligent agent-based simulation of the tragedy of the commons. J Environ Manag 56:159–172

Fox J, Krummerl J, Yaransaran S, Ekasingh M, Podger N (1995) Land use and landscape dynamics in northern Thailand: assessing change in three upland watersheds. Ambio 24:328–334

Gautam AP, Webb EL, Eiumnoh A (2002) GIS assessment of land use/land cover changes associated with community forestry implementation in the middle hills of Nepal. Mt Res Dev 22:63–69

Hjortso CN, Steffen S, Helles F (2006) Applying multi-criteria decision-making to protected areas and buffer zone management: a case study in Royal Chitwan National Park, Nepal. J For Econ 12:91–108

Hurni H (2000) Assessing sustainable land management (SLM). Agric Ecosyst Environ 81:83–92

Ives JD, Messerli B (1989) The Himalayan dilemma: reconciling development and conservation. Routledge, London

Jackson WJ, Tamarkar RM, Hunt S, Shepherd KR (1998) Land use changes in two middle hill districts of Nepal. Mt Res Dev 18:193–212

Jianchu X, Fox J, Vøgler JB, Peifang Z, Yongshou F, Lixin Y, Jie Q, Leisz S (2005) Land-use and land-cover change and farmer vulnerability in Xishuangbanna prefecture in Southwestern China. Environ Manag 36:404–413

Kala CP (2002) Medicinal plants of Indian trans-himalaya: focus on Tibetan use of medicinal resources. Bishen Singh Mahendra Pal Singh, Dehradun, India

Kala CP (2004) Pastoralism, plant conservation and conflicts on proliferation of Himalayan knotweed in high altitude protected areas of western Himalaya, India. Biodivers Conserv 13:985–995

Klevmarken A (1983) There is a need for applied micro-econometrics: survey research about the household. In: Eliasson G, Fries H (eds) Microeconomics: IUI yearbook 1982–1983. Industrial Institute for Economic and Social Research, Stockholm

Lefroy RDB, Bechstedt HD, Rais M (2000) Indicators for sustainable land management based on farmer surveys in Vietnam, Indonesia, and Thailand. Agric Ecosyst Environ 81:137–146

Maikhuri RK, Rao KS, Saxena KG (1996) Traditional crop diversity for sustainable development of central Himalayan agroecosystems. Int J Sustain Dev World Ecol 3:8–31

Maikhuri RK, Nautiyal S, Rao KS, Saxena KG (1998) Medicinal plant cultivation and biosphere reserve management: a case study from Nanda Devi biosphere reserve. Curr Sci 74:157–163

Mallawaarachchi T, Walker PA, Young MD, Smyth RE, Lynch HS, Dudgeon G (1996) GIS-based integrated modeling systems for natural resource management. Agric Syst 50:169–189

Midmore DJ, Jansen HGP, Dumsday RG (1996) Soil erosion and environmental impact of vegetable production in Cameron Highlands, Malaysia. Agric Ecosyst Environ 60:29–46

Naithani HB, Negi JDS, Thapliyal RC, Pokhriyal TC (1992) Valley of flowers: needs for conservation or preservation. Indian Forester 118:371–378

Nautiyal S (1999) Ecosystem function of buffer zone villages of Nanda Devi biosphere reserve. PhD Thesis, H.N.B. Garhwal University, Srinagar, Garhwal (227 pp)

Nautiyal S (2011a) Macro to micro viewpoint of climate change-linking Karnataka to global issue. Nat Resour 2:22–27

Nautiyal S (2011b) Can conservation and development interventions in the Indian central Himalaya ensure environmental sustainability? A socioecological evaluation. Sustain Sci 6:151–167

Nautiyal S, Kaechele H (2009) Natural resource management in a protected area of Indian Himalaya: a modeling approach for anthropogenic interactions on ecosystems. Environ Monit Assess 153:253–271

Nautiyal S, Maikhuri RK, Semwal RL, Rao KS (1998) Conservation through cultivation: a case study of medicinal plants in buffer zone villages of NDBR. In: Research for mountain development. Some initiatives and accomplishments. Himaparyavaran Publication No. 12 Gyanodaya Prakashan, Nainital, pp 342–357

Nautiyal S, Maikhuri RK, Rao KS, Semwal RL, Saxena KG (2002) Agroecosystem function around a Himalayan biosphere reserve. J Environ Syst 29:71–100

Nautiyal S, Maikhuri RK, Rao KS, Saxena KG (2003a) Ethnobotany of the Bhotiya tribe of the buffer zone villages of Nanda Devi biosphere reserve. J Econ Taxonomic Bot 23:119–141

Nautiyal S, Rao KS, Maikhuri RK, Saxena KG (2003b) Transhumant pastoralism and sustainable development: a case study in the buffer zone of the Nanda Devi biosphere reserve, India. Mt Res Dev 23:255–262

Nautiyal S, Rajan KS, Shibasaki R, Rao KS, Maikhuri RK, Bisht IS (2005) Land use intensification and its ecological impacts, in Himalayas of India—lessons for scenario development in mountainous regions. In: Proceedings of 31st international symposium on remote sensing of environment (31st ISRSE), Saint Petersburg, Russia (4 pages), CDROM Publication. http://www.isprs.org/publications/related/ISRSE/

Negi CS, Nautiyal S (2005) Phyto-sociological studies of a traditional reserve forest—Thal Ke Dhar Pithoragarh, Central Himalayas (India). Indian Forester 131:519–534

Nelissen JHM (1991) Household and education projections by means of a microsimulation model. Econ Model 8:480–511

Pilbeam CJ, Tripathi BP, Sherchan DP, Gregory PJ, Gaunt J (2000) Nitrogen balances for households in the mid-hills of Nepal. Agric Ecosyst Environ 79:61–72

Rao KS, Pant R (2001) Land use dynamics and landscape change pattern in a typical microwatershed in the mid elevation zone of central Himalaya, India. Agric Ecosyst Environ 86:113–123

Rao KS, Nautiyal S, Maikhuri RK, Saxena KG (2005) Material and energy flows in village ecosystems with contrasting lifestyles in Nanda Devi biosphere reserve, Central Himalaya, India. J Mt Sci 2:271–293

Sandstrom P, Pahlen TG, Edenius L, Tommervick H, Hagner O, Hemberg L, Olsson H, Baer K, Stenlund T, Brandt LG, Egberth M (2003) Conflicts resolution by participatory management: remote sensing and GIS as tool for communicating land-use needs for reindeer herding in Northern Sweden. Ambio 32:557–567

Sankhayan PL, Gurung N, Sitaula BK, Hofstad O (2003) Bio-economic modeling of land use and forest degradation at watershed level in Nepal. Agric Ecosyst Environ 94:105–116

Saxena KG, Rao KS, Sen KK, Maikhuri RK, Semwal RL (2001) Integrated natural resource management: approaches and lessons from the Himalaya. Conserv Ecol 5:14

Schleiffer R (2005) An intelligent agent model. Eur J Oper Res 166:666–693

Schweik CM, Adhikari K, Pandit KN (1997) Land-cover change and forest institutions: a comparison of two sub-basins in the southern Siwalik Hills of Nepal. Mt Res Dev 17:99–116

Semwal RL, Nautiyal S, Sen KK, Rana U, Maikhuri RK, Rao KS, Saxena KG (2004) Patterns and ecological implications of agricultural land-use changes: a case study from central Himalaya, India. Agric Ecosyst Environ 102:81–92

Sen KK, Semwal RL, Rana U, Nautiyal S, Maikhuri RK, Rao KS, Saxena KG (2002) Patterns and implications for land use/cover change: a case study in Pranmati watershed (Garhwal Himalaya, India). Mt Res Dev 22:56–62

Singh TV, Kaur J (1983) Studies in eco-development: Himalayan mountains and men. Print House, Lucknow, India

Singh JS, Pandey U, Tiwari AK (1984) Man and forests: a central Himalayan case study. Ambio 13:80–87

Thapa GB (1996) Land use, land management and environment in a subsistence mountain economy in Nepal. Agric Ecosyst Environ 57:57–71

Thapa GB, Weber KE (1995) Status and management of watersheds in the Upper Pokhara Valley, Nepal. Environ Manag 19:497–513

Thongmanivong S, Fujita Y, Fox J (2005) Resource use dynamics and land-cover change in Ang Nhai village and Phou Phanang National Reserve forest, Lao PDR. Environ Manag 36:382–393

Virgo KJ, Subba KJ (1994) Land-use change between 1978 and 1990 in Dhankuta district, Koshi Hills, eastern Nepal. Mt Res Dev 14:159–170 von Wiren-Lehr S (2001) Sustainability in agriculture—an evaluation of principal goal-oriented concepts to close the gap between theory and practice. Agric Ecosyst Environ 84:115–129

Waikato E (2003) Coastal values and beach use survey report. Technical Report No. 2003/9. Hamilton New Zealand

Xu JY, Chen LD, Lu YH, Fu BJ (2006) Local people's perceptions as decision support for protected area management in the Wolong Biosphere Reserve of China. J Environ Manage 78:362–372

Chapter 4
Ecosystem Services and Agriculture in Punjab, India

Lalit Kumar and Harpinder Sandhu

4.1 Introduction

Agriculture accounts for 38% of total land-use across the globe (MEA 2005; Sandhu et al. 2015, 2016). Agricultural ecosystems provide humans with food, fibre, and bioenergy etc. which are essential to human well-being. These systems rely on ecosystem services (ES) provided by natural ecosystems like pollination, biological pest control, maintenance of soil structure and fertility, nutrient cycling and hydrological services (Swinton et al. 2007; Wratten et al. 2013; Sandhu et al. 2015). Agroecosystems also produce a variety of ES such as regulation of soil and water quality, carbon sequestration, support for biodiversity and cultural services (Dale and Polasky 2007). Agroecosystems are thus both providers and consumers of ecosystem services. Depending on management practices, agriculture can also be the source of numerous disservices, including loss of forest, nutrient runoff, sedimentation of waterways, greenhouse gas emissions, and pesticide poisoning of humans (Zhang et al. 2007).

4.1.1 Global Food Security, Environment and South Asia

Human population is expected to grow from over 7 billion today to over 9 billion by 2050 (UN 2016). Food demand, especially for processed food, meat, dairy, and fish is expected to grow even faster because of growing urbanization and rising incomes.

L. Kumar (✉)
Department of Business Economics, Bhim Rao Ambedkar College,
Delhi University, Delhi, India
e-mail: lkumar1503@yahoo.co.in

H. Sandhu
School of the Environment, Flinders University, Adelaide, Australia

H. Sandhu (ed.), *Ecosystem Functions and Management*,
DOI 10.1007/978-3-319-53967-6_4

There are huge challenges for the sustainability of food production and the environment to reduce poverty among 800 million people facing hunger and lacking adequate access to food (FAO 2013). More land will be required to grow food crops which are facing competition from conversion to bio fuels to meet increasing energy needs and rapid urbanization. The best lands available for food grain production make up only 12.6% of global land area (16.5 million km^2). Providing adequate food (cereals, oilseeds, etc.) security to the undernourished and reduce poverty at current level would require another 43.7 million km^2 of marginal lands (33.5% of global land area) to be brought under cultivation which would further add to environmental degradation. Tilman et al. (2001) predicts that feeding a population of 9 billion using current methods could result in converting another 0.5 billion hectares of natural habitat to agricultural production, primarily in the developing world. It will also result in doubling or tripling of nitrogen and phosphorous inputs, a twofold increase in water consumption and a threefold increase in pesticide use (Tilman et al. 2011).

A serious limiting factor is expected to be water availability, as 70% of the freshwater used is already devoted to agriculture (Rosegrant et al. 2002). Scenarios prepared by the MEA thus suggest that agricultural production in the future must focus more explicitly on ecologically sustainable management systems (MEA 2005). The countries of the South Asian region (Afghanistan, Bangladesh, India, Nepal, Pakistan and Sri Lanka) apart from their geographical proximity are home to a large section of impoverished population. Almost two thirds (more than 500 million) of undernourished people live in developing countries of Asia, majority in South Asia. This region is characterised by high population, most of which is living at the subsistence level of less than two dollars per day (UNEP 2012).

Share of agricultural sector is falling in the gross domestic product but the population dependent on it remain stubbornly high. Agriculture sector is characterised by growing number of landless labourers, small and marginal farmers, and the scope for intensive agriculture is exhausted (Bhalla et al. 1999; Evenson and Gollin 2003; Mathur et al. 2006). Simultaneously, the region has also become a major producer of grain/cereals and have made huge stride in increasing agriculture production as demonstrated by Green revolution in Indo-Gangetic plains using technological breakthroughs in the agriculture sector (Aggarwal et al. 2000). Green revolution (high yielding varieties, intensive use of input, mechanisation etc.) has transformed Asian agriculture in the last five decades. The technological changes introduced are essentially land-augmenting or land-saving. A big achievement is the involvement of small farm holders in making success of incorporating the new technologies. Asia account for nearly 87% of the small farms with an operational size below 2 ha in the world (out of a total 525 million farms). The importance of the small farm sector in Asia is that it produces 80% of the food consumed in the developing world and feeds one-third of the global population (FAO 2011). The focus has now shifted from extension of cultivation to newer areas to increasing the yield per unit of land available.

This chapter focuses on agroecosystems and how they contribute to food production and provide livelihood to millions of farmers and their families. We take an example of Punjab agriculture and apply ES concept. First, agriculture and its key components in the state of Punjab are discussed. We then elaborate ES produced by agriculture in Punjab. We also discuss and analyze the impacts of current agriculture practices on various ES.

4.2 State of Punjab, India: A Brief Introduction

Punjab is a small state occupying 1.5% of the India's total geographical area. The state supports 2.4% of the country's population with a population density of 484 persons per sq. km. The climate is typically semi-arid and subtropical. Land use in the state is shared by agriculture, forests, water bodies, built up areas, barren and uncultured land. It is a part of the Indo-Gangetic plains formed by alluvial deposits by its two major rivers, the Sutlej and Beas which flows in the central part of the state and Ravi and Ghaggar which touches its northern and southern borders, respectively. This river pattern divides the state into three geographical areas popularly known as 'Majha' (North of Beas), 'Malwa' (South of Sutlej) and 'Doaba' (between Sutlej and Beas). On the basis of geomorphology the state can be divided into hills, intermontane valleys, piedmont plains, alluvial plains, sand dunes, flood plains, wetlands and salt affected areas. Out of total geographical area of 5036 thousand ha, the net cultivated area in the state is 4158 thousand ha while forests cover less than 6%. Agriculture is the largest industry in Punjab and it contributes around 40% to Punjab's GSDP (Gross State Domestic Product) compared to around 20% at national level (Bajwa 2002; Directorate of Economics and Statistics 2016).

4.2.1 Trends in Food Grains Production

In Punjab, from 1980–81 to 2010–11, the total food grain production increased by 129 per, with rice production recording (235% growth) and wheat production (106% growth) during the same period. The proportion of cropped area allocated to food crops increased from 50% in 1960–61 to 88.88% during 2010–11 (Table 4.1). Enhanced productivity and increased area under food crops (rice and wheat) in Punjab has resulted in food production increasing from 3162 thousand tons in 1960–61 to 27,867 thousand tons in 2010–11 (Table 4.2). The production of rice increased from 229 to 16,174 thousand tons and that of wheat increased from 1742 to 16,472 thousand tons over the same period. Rice and wheat accounted for about 62% of the food production in 1960–61, which increased to 98% in 2010–11 (Dhillon et al. 2010).

The cropping pattern in the Punjab has witnessed a seen a major shift in favor of rice-wheat rotation. Agriculture is dominated by rice and wheat crops which now cover over three-quarters of the cropped area and account for 85% of the gross value of crop output. Rice and wheat crops have replaced the traditional crops such as maize, coarse cereals, pulses and oilseeds (Table 4.3). It is also being increasingly grown on all the incremental/marginal land being converted to agriculture.

Table 4.1 Changes in land use pattern and cropping intensity in Punjab (area in thousand hectares)

Particular	1960–61	1970–71	1980–81	1990–91	2000–01	2010–11
Net sown area	3757	4053	4191	4218	4250	4158
Total cropped area	4732	5678	6763	7502	7941	7882
Area under forest	35	123	216	222	280	294
Barren and uncultivable land	–	208	96	83	28	24
Cropping intensity (%)	126	140	161	177	186	191

Source Statistical abstracts of Punjab, Government of Punjab, various issues (2016)

Table 4.2 Production under rice, wheat and other crops in Punjab (000 tonnes)

Year	Rice	Wheat	Maize	Pulses	Total food grain	Oilseeds
1960	229	1742	371	709	3162	121
1970	668	5145	861	308	7305	233
1980	3233	7677	612	204	11,921	187
1990	6506	12,159	335	111	19,223	109
2000	9154	15,551	461	44	25,324	88
2010	10,837	16,472	491	20	27,867	71

Source Statistical abstracts of Punjab, Government of Punjab, various issues (2016)

Table 4.3 Share of various crops in total gross cropped area in Punjab (%)

Year	Rice	Wheat	Pulses	Cotton	Oilseeds	Maize	Sugarcane
1970–71	6.9	40.5	7.3	3.7	5.2	9.8	2.3
1980–81	17.5	41.6	5.0	7.4	3.5	5.6	1.0
1990–91	26.9	43.6	1.9	8.5	1.4	2.5	1.3
2000–01	33.2	43.2	0.7	4.9	1.1	2.1	1.5
2010–11	35.6	44.5	0.3	6.3	0.7	1.7	0.9

Source Statistical abstracts of Punjab, Government of Punjab, various issues (2016)

At present, only American cotton covers about 6% of the gross cropped area in Punjab and all the other crops cover area each with less than 2% of the total cropped area. There is a sharp decline in the proportion of area allocated to pulses from 13% in 1960–61 to 0.30% in 2010–11.

4.2.2 Changes in Land Use Pattern in Punjab

After Green Revolution, barren and uncultivable land, land under non-agricultural use and fallow land were gradually brought under cultivation due to increase in productivity and rising profitability in farming (Bhullar et al. 2006). The net area sown (NAS) increased from 3757 thousand ha in 1960–61 to 4250 thousand ha in

2000–01 (Table 4.1). Since 2000–01, the NAS in the state has declined by about 92 thousand ha, which is largely due to large scale urbanization. There was an increase in the area under forests from 35 to 294 thousand ha during 1960–61 to 2010–11, but the proportion of area under forests is still very small at less than 6% of the geographical area of the state. There was also a simultaneous increase in the total cropped area in Punjab from 4732 to 7882 thousand ha during this period. It implies the increasing intensification of land cultivation as the farmers started growing more than one crop from the same piece of land. Thus, the cropping intensity in Punjab increased from just 126% in 1960–61 to 190% in 2010–11. Currently, Punjab agriculture is characterized by double cropped agriculture system and is one of the most intensive agricultural systems of the country.

4.2.3 Agriculture Growth and Productivity

Assured irrigation, high use of fertilizers, use of high yielding seeds, higher input-use and mechanization are the major reasons for higher productivity growth of food grains in Punjab. There are 434 thousand tractors at present and there is one tractor for every 9.58 ha of net area sown. Almost 98% of the cultivated area is under assured irrigation and 71% of area is irrigated through tube-wells by pumping the ground water resources. The total fertilizer use in the state is 1911 thousand nutrient tons which translates into 460 kg nutrient/ha of net area sown as compared to 144 kg/ha at all India level. Fertilizer use per ha of Gross Cropped Area is highest in India at 243 (kg/ha) (Table 4.4). Almost entire area under the rice and wheat crops is under high yielding varieties. Assured irrigation coupled with high use of fertilizers and high yielding seeds are the major reasons for higher growth.

Green Revolution introduced in the mid-1960s had a major impact on raising yield and output of rice and wheat. By 2000–01, the area under High Yield Variety (HYV) of rice and wheat had touched 95% of total cropped area. At first, the increased input use translated into an increase in productivity of the food crops (Table 4.5). As the inputs use increased exponentially the productivity of rice and wheat more than tripled during 1966–67 to 2010–11. At the same time, there was also significant improvement in case of pulses and oilseeds. Over the period, the productivity of maize and other cereals has also shown an increase from 1383 to 3693 kg/ha and from 848 to 3652 kg/ha, respectively (Table 4.5).

4.2.4 Distribution of Operational Land Holding in Punjab

Two-thirds of the state's population lives in rural areas and mostly relies on agriculture for its livelihood. The data on operational land holdings since 1970–71 indicates that the number and proportion of small and marginal holdings are

Table 4.4 Some input use indicators in Punjab agriculture

Indicators	1970	2010
Area under HYV seeds (% of total cropped area)	69	98
Gross cropped area (ha)	4732	7882
Gross irrigated area as % of gross cropped area	75	98
Area irrigated by canal (% of total irrigated area)	44.7	25.9
Area irrigated by tube wells (% of total irrigated area)	55.1	72.6
Fertilizer consumption (NPK) (000 nutrient tons)	213	1911
Number of tractors	45,535	367,107

Source Statistical abstracts of Punjab, Government of Punjab, various issues (2016)

Table 4.5 Average productivity of food grains in Punjab (kg/ha)

Decade	Rice	Wheat	Maize	Other cereals	Oilseeds
1960–70	1000	1240	1384	815	769
1970–80	2155	2305	1555	1176	912
1980–90	3145	3283	1467	883	902
1990–2000	3433	3920	1602	1107	821
2001–02	3517	4532	1585	1043	893
2004–05	3917	4221	1786	893	887
2006–07	3868	4210	2762	1045	904
2009–10	4010	4307	3693	900	850

Source Compiled from various issues of Punjab statistical abstract, Government of Punjab (2016)

decreasing in Punjab. The number of marginal holdings (with less than one hectare land) in 1970–71 was 517,568 (37.36% of total holdings); this declined to 122,706 (12.31% of total holdings) in 2000–01. Similarly, the number of small holdings (with 1–2 ha of land) declined from 260,083 (18.91% of total holdings) in 1970–71 to 173,071 (17.35% of total holdings) in 2000–01 (Table 4.6). As agriculture in post-green revolution has become dependent on capital-intensive technology, it requires large investment both in fixed as well as working capital. Therefore, marginal and small holdings are fast losing their viability.

Planning Commission of India estimates that 16% of the state population was below poverty line in the mid-1980s. The proportion of marginal farmers below the poverty line increased to 34.07% by 1990–91. The agrarian situation toward the later part of 1980s and the entire period of 1990s showed evidence of a steady deterioration in the situation of small and marginal farmers.

Table 4.6 Distribution of operational land holding in Punjab

Size class (ha)		1970–71	1980–81	1990–91	2000–01
Marginal (0–1)	% of total	37.36	19.21	26.50	12.31
Small (1–2)	% of total	18.91	19.41	18.24	17.35
Medium (2–4)	% of total	20.44	27.99	25.85	32.91
Large (4–10)	% of total	18.02	26.20	23.40	30.18
Extra large (>10)	% of total	5.00	7.19	6.01	7.25

Source Compiled from various issues of Punjab statistical abstract, Government of Punjab (2016)

4.2.5 Decomposition of Agricultural Growth in Punjab

The annual growth in aggregate value of crop output was 5.94% in the initial green revolution period (1967–80), with the contribution of area under cultivation being 38% and aggregate crop yield effect was 40.9% during this period. The decomposition of changes in the production of rice, wheat and total food grains in Punjab shows that the contributions of area towards overall food production in Punjab have increased over time and the contributions of yield and the interactions between area and yield together have played a relatively less important role. While the incremental yield increase (along with interaction with area) played more significant role in enhancing rice production during 1960–61 to 1980–81, area played more important role during 1980–81 to 2000–01, and the yield emerged again as an important contributor during the last decade (Table 4.7).

As new technology spread during the post-green revolution period from 1981 to 1990, the effect of crop yield has doubled to that of area. Moreover, during the stagnation period post-1990s, growth was small and largely contributed by area expansion (59%); the effect of yield decreased to less than one-third indicating technology fatigue. This is indicative of the fact that future growth in agriculture with the present crop patterns and technology will come largely from the area expansion which is limited due to water constraints and already high levels of cropping intensity of 190%.

4.2.6 Stagnation in Agricultural Growth

Agricultural growth rate in Punjab has slowed down from 5.0% in 1980s to 2.4% in 1990s and to 1.9% in 2000s (Johl 1986, 2002). The yield of rice has almost stagnated in Punjab, going up only by 0.02% annually in the 1990s and wheat has slowed significantly from 2.96% annual gain in 1980s to 1.96% in the 1990s compared to all India average of 3.2%. Overall, the crop sector grew by 1.3% per annum in the 1990's, less than 1% in the 2000's as compared to 4.8% per annum in the eighties. In the last five years, the production of food grains in Punjab has increased by only 2% as against 8.6% rise in its population. As a result contribution

Table 4.7 Decomposition of agricultural growth in Punjab

Year	Agriculture growth (%)	Constituent of agricultural growth	
		Area effect (%)	Land productivity effect (%)
1967–1981	5.94	37.54	62.46
1981–1991	5.16	28.92	71.17
1991–2001	0.83	59.04	40.96

Source Compiled from Singh and Sidhu (2004)

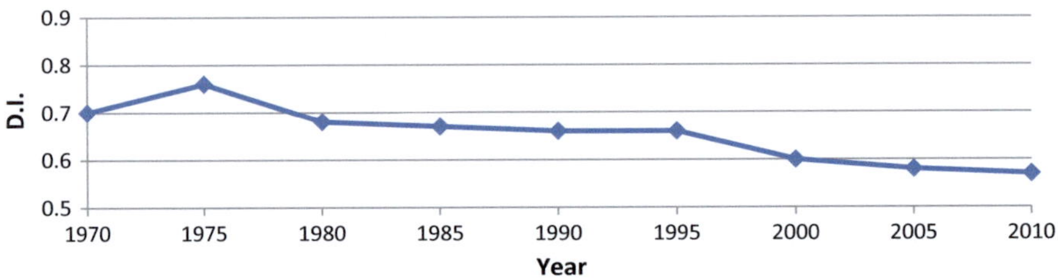

Fig. 4.1 Diversification index (D.I.) in Punjab. *Source* Compiled from various issues of Punjab statistical abstract, Government of Punjab (2016)

of agriculture sector to the state's GDP has declined from 46.13% in 1993–94 to 37.07% in 2004–05. For wheat the highest yield of 4532 kg/ha was obtained during the year 2000–2001 (Table 4.5). Thereafter, it has fluctuated between 4210 and 4367 kg/ha.

4.2.7 Declining Crop Diversity in Punjab

As the area under rice and wheat has increased after green revolution, the crop variety has declined drastically (Singh and Sidhu 2004). The combined area under rice and wheat crops which was less than 40% during late-1960s and has touched recently to more than 80%. The Diversification Index (DI) for the state declined from 0.707 in 1970–71 to 0.591 in 2001–02 (Fig. 4.1). This has been at the expense of area under other Rabi crops especially gram, barley, rapeseed, mustard and sunflower. Area under total pulses has also reduced sharply. Area under cotton has been fluctuating around 5–6% of total gross copped area of Punjab. In place of hardened and time tested broad range of traditional varieties which were naturally suited to the climatic conditions of Punjab, farmers have adopted a narrow range of High Yield Varieties. The area under input intensive HYVs of wheat has increased from 69% in 1970–71 to 100% in 2000–01. This highlights the fact that rice and wheat have continued to grow in terms of area and production at the cost of other crops, with the diversity in the output mix decreasing continuously over time—signaling a decline in genetic diversity, which is an important ecosystem service.

4.2.8 Loss in Biodiversity of Indigenous Crops Varieties

Before green revolution, 41 varieties of wheat, 37 varieties of rice, four varieties of maize, three varieties of bajra, 16 varieties of sugarcane, 19 species/varieties of pulses, 9 species/varieties of oil seeds and 10 varieties of cotton were grown in Punjab and propagated through pure line selection by farmers. However, currently the number of varieties in use by farmers has decreased to 13 varieties of wheat, 8 varieties of rice, 4 varieties of Basmati rice and 4 varieties of pulses (Johl and Ray 2002). Punjab Agricultural University has found that more than 87% of the area under wheat is currently using a single high yielding variety PBW 343 (released in 1995). Similarly, high yielding rice varieties of only PR series-106, 108, 111, 113, 116, 118 are grown throughout the state in every season. Thus, food production is dependent on the narrow and alien genetic base and the natural ecologically adapted bio diverse agricultural base is virtually lost.

4.2.9 Total Factor Productivity (TFP) and Agricultural Growth

The analysis of total factor productivity (TFP) growth of major crops in Punjab shows that the technological gains of early years of green revolution have exhausted their potential (Table 4.8). Agriculture has been experiencing diminishing returns to input use and a significant proportion of the gross cropped area has been facing stagnation or negative growth in TFP. The share of TFP in total output has also been declining and negative in some cases, which implies that the growth in output is more because of the increased use of inputs rather than the technology factor. Singh and Hossain (2002) showed that growth in total factor productivity (TFP) for rice was negative (−1.77% per annum) during the period 1990–91 to 1997–98, with the component of technical change occurring at a very slow pace (0.89% per annum). In the case of wheat, although TFP growth was positive (1.24% per annum), the growth in technical change was slow (1.01% per annum) during the period 1990–91 to 1997–99. For both rice and wheat, environmental degradation (sustainability) was found to contribute negatively to TFP, with the negative contribution being much larger in paddy (−5.04% per annum) than in wheat (−1.58% per annum). Kumar (2002) showed that TFP growth in Punjab has been negligible in the 1990s (0.05% per annum) as compared to the 1980s (1.55% per annum). Deceleration in TFP in the Punjab was mainly due to intensive input use and to the increasing occurrence of weeds, insects and disease in the continuous mono cropping of food grains.

As food grain productivity declines, decomposition of the impact of various factors on TFP growth show that area extension accounted for about 45% of the TFP growth, public research (36%), literacy (10%), infrastructure (8%), and urbanization (1.5%).

4.3 Ecosystem Services in Punjab Agriculture

Agricultural ecosystems rely on variety of supporting ecosystem services to provide food, fibre and bio-energy, pharmaceuticals as well as a range of accompanying but non-marketed ecosystem services (Box 4.1). ES from agriculture include regulation of water and climate systems, aesthetic and cultural services, as well as enhanced supporting services (such as soil fertility). They are a major contributor to agricultural productivity; leading to increases in agricultural crop yield and thereby contributing towards food security. At the same time, agriculture produces more than just crops. Current agricultural practices have environmental impacts that affect a wide range of ecosystem services, including water quality, pollination, nutrient cycling, soil retention, carbon sequestration and biodiversity conservation. Thus there is a strong linkage between ES, agriculture and food security.

The following section looks at the trends of the key ES, corresponding drivers of change and their impacts on agricultural practices in Punjab:

- Water quantity and quality.
- Soil structure and fertility.
- Nutrient cycling.

4.4 Water Quantity and Quality in Punjab

4.4.1 Water Use and Availability Scenario in Punjab

In Punjab, almost 97% of the cultivated area is under assured irrigation with nearly 80% of the water resources being used for agricultural production. Water is a major input under the Green Revolution strategy which has created an increasing demand for irrigation. In 1950–51, 25–30% of the cropped area in Punjab was irrigated

Table 4.8 Total factor productivity (TFP) growth in Punjab (per cent per year)

Period	Output growth	Input growth	TFP growth
Early green revolution (1970–1985)	4.72	1.10	3.62
Late green revolution (1970–1985)	−0.92	−0.12	−0.79
Overall green revolution (1970–2000)	1.67	0.44	1.23

Source Singh and Hossain (2002)

Provisioning services
These include food and services for human consumption, ranging from food, fibre, fire wood, livestock, raw materials, genetic resources, ornamental resources and medicinal resources.

Regulating services
These include ecological processes such as gas regulation, climate regulation, disturbance regulation, water regulation, water supply, erosion control and sediment retention.

Cultural services
Agriculture provides cultural services such as aesthetics, recreation through conserving field-boundary vegetation or by planting native trees.

Supporting services
Supporting ecosystem services are pollination, biological control, carbon accumulation, mineralization of plant nutrients, soil formation, nitrogen fixation.

Box 4.1 Four types of ecosystem services in agricultural ecosystems (*Photos* by H. Sandhu)

against 17.11% at the national level which increased to about 90% in 1983–84 as compared to 28.61% at national level. The Green Revolution increased the need for irrigation water at two levels. Firstly, the shifts from millets, pulses and oilseeds to wheat and rice cycle increased the need of water throughout the year. Secondly, the new hybrid varieties of wheat and rice also increased the intensity of irrigation. High yielding varieties of wheat and rice needed more irrigation than traditional varieties. In 1970–71, the area under canal irrigation was 12.86 lakh ha which rose to 16.6 lakh ha in 1990–91. It fell to 9.62 lakh ha in 2000–01 At present, gross water requirements for the state are estimated at 6.15 M ha-m (million hectare per annum), against the current availability of only 3.66 M ha-m, comprising 1.52 M ha-m of surface water and 2.14 M ha-m of groundwater resources.

Out of the total water availability, 0.35 M ha m is lost during conveyance/transportation and only 1.45 m ha m is available at the outlet that irrigates about 1.0 million hectares of land. The remaining 3.24 million hectare of land is irrigated with the use of groundwater. The total sustainable availability of groundwater is 1.68 M ha-m. Therefore, the total water availability including surface as well as groundwater is 3.13 M ha-m. The current crop production pattern dominated by wheat-rice crop rotation requires 4.37 M ha-m of crop irrigation water per annum (Government of Punjab 2011). This implies a gap of 1.24 M ha-m between demand and supply which is met through overutilization of groundwater resources.

4.4.2 Ground Water Status in Punjab

According to the Central Ground Water Board, the annual water discharge in Punjab is 23.78 billion cubic meters and the net water availability is 21.44 billion cubic meters. The withdrawal of groundwater for irrigation purpose is 31.16 billion cubic meters annually, which is 145% of the availability. This is causing fast decline in the water table which has been falling at the rate of 50–55 cm per annum. In 1973, the area where the water level was below 10 meters stood at 3%. In 2008 it was estimated to be more than 75%. The proportion of area irrigated through groundwater resources in Punjab increased from 41.04% in 1960–61 to 71.77% in 2006–07. Such proportion increased from 70.96 to 88.71% in the Central Zone of Punjab which has been most severely affected (Statistical Abstract of Punjab, various issues)[1]. The dominance of rice-wheat crop rotation has converted the Punjab state from a water-surplus to a water-scarce state. At present, gross water requirements for the state are estimated at 6.15 M ha-m, against the current availability of only 3.66 M ha-m, comprising 1.52 M ha-m of surface water and 2.14 M ha-m of groundwater resources. It indicates a total deficit of about 2.49 M ha-m of water—a case of severe water imbalance in the state.

4.4.3 Surface Irrigation Status in Punjab

The area irrigated by canals expanded from 1292 thousand ha in 1970–71 to 1669 thousand ha in 1990–91 and then declined to 1116 thousand ha in 2010–11. On the

[1]Office the Statistical Adviser, Economic and Statistical Organisation, Government of Punjab. Statistical abstract of Punjab, various issues, Chandigarh.

other hand, the area irrigated by tube wells increased from 1591 thousand ha in 1970–71 to 2954 thousand ha in 2010–11. While the canals accounted for almost 45% of the irrigated area in 1970–71, such proportion is just 27% in 2010–11. Almost 73% of the cultivated area is currently being irrigated through groundwater, which has serious implications for the sustainability of the groundwater resources in Punjab. The number of tube wells increased from 192 thousand in 1970–71 to 1382 thousand in 2010–11. The tubewell density i.e. the number of tube wells per thousand ha was just 32 which increased to 67 in 1980–81 and further to 209 in 2004–05. On an average there are 28 tube wells per sq. km. of net sown area in Punjab alone. At the same time the growth in irrigated area fell to negative 0.02% after 2000's. At present, 11.01 lakh hectares are under canal irrigation which is around 36% of the total irrigated area in the state. The state has 86% cropped area and 98% of this is under irrigation that uses nearly 84% of the state's water resources. Out of this, rice consumes 34%, wheat 30% and other crops 36%. The area under canal irrigation is declining every year because of decreasing carrying capacity of canals and less water coming from the rivers of Sutlej, Beas and Ravi which fell from 17.17 million acres feet (MAF) in 1990 to around 14 MAF at present.

4.4.4 Water Demand and Supply Status in Punjab

In order to increase the food grain production in Punjab, the area under food grain drastically increased from 3.06 million hectares in 1960–61 to 6.51 million hectares, a growth of 112%. The area under rice increased at a growth of 1147% during the given period. It increased from 0.23 million hectares in 1960–61 to 2.83 million hectares in 2010–11. This increase in area under rice has in turn increased the water requirements in the state for irrigation and also increased its contribution in the state water demand. The contribution of rice in total water demand has increased from 5% in 1960–61 to 38% in 2010–11. Thus the water demand has increased to 4.8 million hectares meter in 2010–11. With supply more or less static around 3.13 million hectares meter, the state is witnessing a water deficit of 1.63 million hectares meter per annum.

The data in Table 4.9 clearly shows the correlation between the increase in area under rice and wheat and the growing water deficit in the state. The water deficit has grown from 0.66 M-ha-m in early 1980's to 1.67 M-ha-m by 2010. This increasing deficit is being met by overdrawing water from the ground which is depleting at an alarming rate. As per the findings of Soil Science Department of Punjab, Agricultural University, Ludhiana, the level of ground water is going down by 50–55 cm in Punjab every year since 1993–94. If this situation continues it will lead to the level of water going further down by more than 70 feet in the 50% region of Punjab, 100 feet in the 14% and more than 130 feet in 2% area of Punjab by 2013. In 1964, the water level was at 5 feet in the 23% region of the state, 5–10 feet in the

Table 4.9 Water demand and supply (M-ha-m)

Year	1960–61	1970–71	1980–81	1990–91	2001–01	2010–11
Demand	2.7	3.12	3.79	4.43	4.74	4.80
Supply	3.13	3.12	3.13	3.13	3.13	3.13
Deficit	0.43	–	−0.66	−1.3	−1.61	−1.67
Area under food grain production (000 ha)						
Rice	0.23	0.39	1.18	2.02	2.61	2.83
Wheat	1.40	2.3	2.81	3.27	3.41	3.51
Total	3.06	3.93	4.85	5.68	6.28	6.51

Source Estimated by author

Table 4.10 Pattern of annual average rainfall in Punjab (in mm)

Year	1970	1980	1990	2000	2004	2008	2009	2010	2013	2014
Rainfall	672	739	754	391	392	529	384	472	619	385

Source Indian Meteorological Department, 2016

20% region; 10–20 feet in 11% of the region and the remaining 42% area of Punjab had the water level of 25 feet.

The increased demand of water, reduction in canal capacity due to siltation and the easy credit facilities for tubewell installation with state subsidy, besides liberal facilities for electrification of tube wells have led to over exploitation of ground water for irrigation purpose. The centrifugal pumps are becoming redundant and being replaced with submersible pumps especially in central Punjab due to depletion of underground water table. The availability of river water has come down because the annual average rainfall in the region has gone down substantially (Table 4.10).

The World Watch Institute has warned that if the water is used at this rate, the whole ground water in Punjab shall be finished by 2025. Climatic changes and decline in rain fall has further added to water problem of the state in the form of inadequate recharging of ground water. The annual rain fall which was 672.3 mm in 1970 rose to 754.6 mm in 1990. It has since been falling rapidly and declined to 391.9 mm in 2000. In the new century, it has remained below the mark of 400 mm and stood at 375.5 mm in 2004 and at 384 mm in 2009. Scientific studies in Punjab have indicated that the current use and withdrawal of water resource can sustain 1.6 mha of land under rice and 1.2 mha areas should be brought under other crops for sustaining rice productivity and maintaining the ecological balance.

4.4.5 Waterlogging and Salinization in Punjab

Water logging causes depletion of oxygen and increase of carbon dioxide in the root zone of crops which causes loss of plant nutrients and the loss of useful microorganisms at the expense of the growth of harmful ones. It also causes chemical degradation due to accumulation of salts at the soil surface leading to an ecological imbalance. All these factors result in reduced or near zero productivity.

The twin problems of waterlogging and salinization in South-west Punjab are broadly due to the digressional location of the area along with the lack of proper drainage system. The area has low and erratic rainfall of 100–400 mm per year (Hira et al. 1998). The ground water is brackish with high EC (2–12.6 d sm^{-1}) and RSC (3–35.1 meq^{-1}). While groundwater is declining at an alarming rate in high yielding fresh groundwater regions in northern and central districts, the south-western parts of Punjab are facing problems of severe water logging and salinization.

The South-Western districts comprises 34% area of Punjab and out of 17.2 lakh hectare areas of the zone, 6.3 lakh hectare areas has a rising water logging problem. Excessive utilization of water mainly for rice in south west Punjab and inadequate drainage system are the major causes of water logging. Water logging has destabilized water cycle of the area by adding more water than its natural drainage potential. This has led to desertification and salinization of the land. The water table of this area is rising at the rate of 0.5–1 m per year. The cotton is very sensitive crop to water logging and cannot tolerant excess water in its root zone. The original cotton/bajra/maize-wheat/gram system has been replaced by rice-wheat crop rotation by leveling of extensive sand dunes and conversion of these to irrigated fields. Changing cropping pattern, aridity, rise in water level, use of poor quality irrigation water and canal seepage in both the Rajasthan feeder and Sirhind canal have compounded the problem critically.

With the rise of water table in the South-Western Punjab, the farmers have now taken to rice cultivation. Cotton has been replaced by rice and as rice cultivation needs more water, the farmers are demanding more canal water because the underground water in this area is brackish, which is unfit for the irrigation purposes. Water logging has ruined the fertility of land and a marginal type of well water gives 200–400 kg of salt per hectare during one season. The increased water logging has not only affected the cotton crop of this region but the stagnation of water for long time has increased the humidity of the environment. It has increased insect/pest in the environment which is proving resistant to recommended insecticide sprays.

4.5 Soil Structure and Fertility

Maintenance of soil fertility is essential to sustain agricultural production. A healthy soil provides for the following ecosystem function to agriculture: ensure proper retention and release of water and nutrients, promote and sustain root growth, maintain soil habitat, respond to management and resist degradation. While intensification may increase soil fertility in the short run, many researchers have identified numerous ways that intensification causes soil degradation.

The introduction of machinery compacts soil reducing its capacity to absorb rainwater, thus speeding up runoff and resulting in erosion. This is also known as "mining" of the soil's organic matter and nutrients because they are removed rather than renewed with use. Further, the "green revolution" model of intensive agriculture is highly water intensive and requires increased use of chemical fertilizers and pesticides. The increased irrigation necessary for high-yield plant varieties causes soil salinization and waterlogging. Also the application of chemical fertilizers can cause excesses of soil nitrogen. In Punjab, high inputs of nitrogen fertilizers increase pesticide use. Nitrogen encourages growth of weeds better adapted to high levels of nitrogen. In addition, nitrogen can lead to greater damage from animal pests, because plants become more nutritious for herbivores as nitrogen level increases.

The soils of Punjab are low in Nitrogen (N) content, low to medium in Phosphorus (P) and medium to high in Potassium (K) (Department of Agriculture, Punjab) (Table 4.11). The rice wheat cropping pattern in Punjab has resulted in a break in the re-cycling of nutrient in two ways. First, the HYV led to reduction of organic material available for recycling into soil which increased the grain yield at the cost of straw. The traditional rice varieties had four to five times as much straw as grain. High yielding rice produces a one to one ratio of grain to straw. The scarcity of straw reduced biomass availability for fodder and mulch. This resulted in a break in nutrient re-cycling. Secondly, it demanded higher nutrients which caused toxic chemicals in the soil and micronutrient deficiencies. Rice wheat sequence that yields 7 tons/ha of rice and 5 tons/ha of wheat removes more than 300 kg N, 30 kg P and 300 kg of K per hectare per annum from the soil. Consequently, partial factor productivity of NPK application in Punjab has dropped from 80.9 in 1966–67 to 16.0 in 2003–04.

4.5.1 Soil Organic Carbon (SOC)

SOC is taken as an index of plant available Nitrogen (N) since it is related with N supplying capacity of soil (Table 4.11). SOC has a significant effect on all soil properties related to its fertility potential. Organic carbon content has declined to low levels because of very low or limited application of organic manures and non-recycling of crop residues. Loss in organic carbon means wasteful extra

Table 4.11 Trends in available organic carbon content in soils of Punjab (1981–2001) (percentage sample)

Year	Soil organic carbon (N)			Phosphorus (P)			Potassium (K)		
	High	Median	Low	High	Median	Low	High	Median	Low
1981–91	0	20	80	48	29	23	08	45	46
1991–2001	5	25	70	62	33	15	09	51	40
Current status	12	32	57	36	18	46	08	–	93

Source Benbi and Brar (2009)

application of chemical fertilizers, loss in soil biological activity and poor soil moisture retention.

4.5.2 Destruction of Soil Fertility

Soil fertility declines are well recorded for Punjab. Numerous long-term experiments have shown a declining trend in productivity even with higher application of N, P, and K fertilizers and the use of modern intensive farming.

More nutrients have been removed than added through fertilizers to the fields and farmers are applying more fertilizers to get the same yield than they were getting with less fertilizer 20–30 years ago. Micro-nutrient deficiencies have started appearing with the adoption and spread of intensive agriculture. Zinc deficiency has become most widespread in the entire region. The average organic content of the soil of the state has gone down from 0.5 from early green revolution period to 0.2% at present. Analysis by Punjab Agriculture University during 1981–1992 reported that 78% soils of Punjab were low (<0.4% SOC) in organic carbon, 21% medium (0.4–0.75% SOC) and 0.5% high (>75% SOC) in soil organic carbon. In 1950–1960, the soil of Punjab was deficient only in nitrogen, but with the adoption of the new agricultural strategy, it is short of all other micro and macro nutrients like Potassium, Zinc and Phosphorous. During the eighties, the deficiency of Manganese and Sulfur were also seen. In the era of nineties, the soil became poor in Copper and other nutrients.

A comprehensive study by Benbi and Brar (2009) on the level of SOC in Punjab has shown, unexpectedly that intensive agriculture has resulted in improvement in soil organic carbon (SOC) status. As a weighted average for the whole state, SOC increased from 2.9 g kg^{-1} in 1981/82 to 4.0 g kg^{-1} in 2005/06, an increase of 38%. Increased productivity of rice and wheat resulted in enhanced Carbon sequestration in the plough layer by 0.8 t/ha per ton of increased grain production. Soil pH declined by 0.8 pH units from 8.5 in 1981/82 to 7.7 in 2005/06. This pH decline has positive implications for availability of phosphorus and micronutrients such as Zn, Fe and Mn. The status of available P in soils increased from 19.9 kg ha^{-1} in 1981/82 to 29.2 kg ha^{-1} during 2005/06. The status of plant-available K in soil remained almost unaltered and averaged 106 and 123 mg kg^{-1} soil in 1981/82 and 2005/06, respectively. The analysis showed that intensive cultivation of a rice-wheat system unexpectedly resulted in improved C sequestration, a favorable pH environment and improvement of the soil salinity.

Certain management practices such as conservation tillage, no-till practices can help address these issues of detrimental impacts of the current soil management practices (Abrol and Sangar 2006). Conservation or no-till allows planting of new crop in the residues of previous year crop. It can minimise surface run-off and help build soil structure by increasing soil organic matter.

4.5.3 Nutrient Cycling

Nutrient management is the most important factor in sustainable crop productivity as the nutrients account for at least 35% of plant growth and yield. Studies shows that rice-wheat production system at 8–9 tons/ha production level remove about 625–675 kg/ha of NPK annualy.

The addition of nutrients through chemicals and fertilizers does not match the crop removal resulting in nutrient mining of soils. Energy use is high in intensive cereal production systems. Of the total energy used for crop production, fertilizer and chemical energy inputs comprise 47% for wheat and 43% for rice (Khan et al. 2009). About 60% of this is due to nitrogen (N) fertilizers alone. Wheat and rice are the most nutrient exhaustive crops of the state removing 80% of N, P and K, 80% of S and 77% of Zn of the total removal of nutrients per year. Based on nutrient requirements for each crop based on their productivity levels, it was estimated that the addition of N through chemical fertilizers was higher than removal since 1980–81 and this positive difference between addition and removal has been increasing over years resulting into positive N balance. In many areas of the state, the application of nitrogen is higher than the recommended doses especially for wheat and rice crops. This has led to unsustainable rise in fertilizer use ratio of 35:9:1 as compared to recommended usage ratio of 4:2:1 (Yadvinder-Singh et al. 2005).

4.6 Trade-Offs in ES in Agriculture

4.6.1 Yield Gap of Productivity in Punjab

Potential yield can be interpreted as the upper limit that can be achieved by the current varieties in the given production system. A review of data from experiments, long-term field trials and simulation studies has shown potential yields of rice-wheat in the optimal rice-wheat planting system to be around 18.29 tons/ha in Punjab. The actual productivity data for Punjab were correlated with the nutrient consumption data and has revealed a good linear correlation ($R^2 = 0.78$). The simple linear regression analysis shows the following result: $Y = 16.709X + 143.35$ where Y is wheat yield (kg/ha) and X is fertilizer consumption (kg/ha) (Majumdar et al. 2013). This large yield gaps in Punjab are due to inappropriate nutrient management and wheat yield response ranged from 500–4750 kg ha^{-1}, 67–2806 kg ha^{-1} and 0–2222 kg ha^{-1} for N, P and K respectively, which is related to the soil nutrient supplying capacity. Khurana et al. (2008) working on irrigated wheat at 56 sites in six main wheat producing regions of Punjab also highlighted the variability in soil fertility across farmers' field sites. Fertilizer use efficiency in rice is only 30–35% whose overuse/misuse causes environmental and ecosystem deterioration (Yadvinder-Singh et al. 2004). Due to high field to field variability, the possibility of over or under-application of nutrients are very high with its economic and environmental consequences. This leads to inefficient use of added nutrients, as application rates do not consider the spatial variability in nutrient requirements among the fields. Zinc and other micronutrients are not applied in adequate amounts to prevent increasing deficiencies of these nutrients. Sanyal et al. (2010) also pointed out the lack of potassium application as one of reasons for yield stagnation in the rice-wheat system in the state.

Fertilizer (N) use constitutes the major component of the total fertilizer use pattern. Even under the best practices, not more than two third of the N added as a fertilizer can be accounted for as being utilized by crop plants or to be present in the soil. Field experts have recommended application of fertilizer (N) at 120 kg N/ha for rice-wheat cropping pattern in Punjab but in practice more than 200–300 kg N/ha are used in the fields in efforts to overcome the low fertilizer N use efficiency (35–40%) (Yadvinder-Singh et al. 2009). Nitrate content in ground water is continuously increasing and it has been linked to the inefficient fertilizer N use in the rice-wheat system. To obtain the projected rice yield of 8.0 Mt/ha by the year 2025 in Punjab, it would become necessary to apply 280 kg N/ha at 33% fertilizer N recovery efficiency (Cassman and Pingali 1995).

There has been a built-up of nitrogen in the Punjab soils over time as the proportion of area with low levels nitrogenous contents declined from 78 to 32% over time. Punjab has the highest per hectare usage of fertilizers (NPK) at 237.05 kg/ha in 2009–10, as against the all India average of 135.27 kg/ha in the same year. High usage of nitrogenous fertilizers with relative under-utilization of other fertilizers and micronutrients has increased imbalance of micro nutrient in the Punjab soils.

4.6.2 Emergence of New Pest and Diseases

On an average, rice farmers in Punjab overuse pesticides by more than 30% and over application of nitrogen fertilizer is partially responsible for the overuse of pesticides. An increase in the cropping intensity, use of fertilizers and intensive irrigation has changed the ecological balance in favour of weeds and it has become difficult to control weeds manually. Before Green Revolution, indigenous varieties were resistant to local pests and diseases. If the crop was affected by certain diseases, it affected only some parts of the crop while others had the resistance to survive. Crop rotations also helped in pest control. Many pests were specific to particular plants. Planting different crops in different seasons and different years reduced the pest population. The rice crop was considered to be comparatively free from insects/pests before the Green Revolution. The rice cultivation in Punjab is now affected by near about 4 insects and 12 diseases. The rice leaf folder was first recorded in 1964 and now it present in all rice growing areas of the Punjab. The yellow stem borer, white backed plant hopper, green leaf hopper, rice thrips are the new pests-insects created by the Green Revolution. Among the diseases to which the rice cultivation is affected is brown spot, false smut and sheath rot. The new high yielding dwarf seed varieties of wheat have turned minor diseases such as Karnal Bunt into epidemic form. Other diseases like Leaf Rust, Loose Smut, and Stripe Rust has affected the new wheat varieties such as PV 18, Kalyan 227 etc. The weeds have also created a serious threat to paddy-wheat rotation in the Punjab. Problem of *Leptochloa chinensis* (*chini ghaas*) and *Ischameum rugosum* (*kanaki*), which were earlier hard to see in rice crop, can be seen everywhere in rice crop (Mahajan et al. 2009). Also broad leaf weeds (*Gharilla*) have also started appearing in rice and cannot be killed with the common used herbicide (butachlor).

The use of insecticide and pesticides has increased from 624 tons during 1960–61 to 6900 tons in 2003–04. The per hectare use of chemical pesticides has increased from 0.132 to 0.873 kg during the same period. The use of weedicides has also increased and Punjab accounts for 60% of the total weedicides consumption in India. More than 90% growers of rice and wheat use weedicides. The indiscriminate use of pesticides has led to plants developing resistance to pesticides and insecticides. Food grains have found to be contaminated with DDT and HCH residues 4 ppm which is above the recommended maximum residue limit of 0.1 ppm. In 2005 Punjab Pollution Control Board study pointed to a cancer-pesticide link. Samples collected from villages in Punjab had pesticide content in the blood which included aldrin (80%), DDT (50%) and endosulfin (24%) among others. Diseases ranged from cancer, lymphoma, genetic changes to brain toxicity and birth defects (Times of India, August 17, 2011). At 90 cancer patients for every 1 lakh population in Punjab, it's more than the national of 80 per lakh. The Malwa region of the state—known as the cancer belt—has the highest average of 136 cancer patients per 1 lakh people, according to data provided by Punjab government. Data over the last five years has shown that, on an average, 18 people die of cancer each day. The excessive use of chemical fertilizer like urea, DPK, NPK, etc. and pesticides (insecticides and weedicides) in Punjab have resulted in the disorders of endocrine glands e.g., thyroid, parathyroid, pituitary, kidneys and adrenals (Kheti Virasat Mission 2016).

4.6.3 Loss of Soil Fertility Due to Rice Residue/Straw Burning and Environmental Damage

Crop residues are an important component of soil fertility management. They are currently burnt, especially rice residues leading to degradation of natural resources especially soil. Rice residues can be converted to high-value manure of a better quality than farmyard manure, and their use, along with chemical fertilisers, can help sustain or even increase yield (Sidhu et al. 1998). The residues of rice crop from one hectare land on composting give about 3 tons of manure as rich in nutrients as farmyard manure.

Crop residues are good sources of plant nutrients and are important components for the stability of agricultural ecosystems. In Punjab where mechanical harvesting is practiced, a large quantity of crop residues are left in the field which can be recycled for nutrient supply. Farmers remove wheat straw for feeding the animals but rice straw due to its high lignin and silica and low protein content is not suitable for this purpose (Pathak et al. 2006). Short turn-around time available for rice residues disposal interferes with tillage and seeding operations for subsequent wheat crop. This compels farmer towards burning the rice residue. Crop residues retention in the soil improves soil physical (e.g., structure, infiltration rate, plant available water capacity), chemical (e.g., nutrient cycling), and biological (e.g., Carbon sequestration). The critical level of Carbon input requirement for maintaining SOC at the required level has been calculated as 2.47 t/ha/year for rice-based systems (Srinivasarao et al. 2013). A rice wheat sequence that yield 7 tons/ha of rice and 4 tons/ha wheat removes more than 300 kg N, 30 kg P and 300 kg K/ha from the soil. About 25% of nitrogen (N) and phosphorus (P), 50% of sulphur (S), and 75% of potassium (K) uptake by cereal crops are retained in crop residues, making them valuable nutrient sources. One ton of wheat residue contains 4–5 kg N, 0.7–0.9 kg P, and 9–11 kg K (Yadvinder-Singh et al. 2005). N, P and K amounts in rice straw is 6.2 kg N, 1.1 kg P and 18.9 kg K per ton of straw. Besides NPK, one ton of rice and wheat residues contain about 9–11 kg S, 100 g Zn, 777 g Fe and 745 g Mn. Crop residue play an important role in the cycling of nutrients despite the dominant role of chemical fertilizers in crop production. Considering 90% of rice straw and 30% of wheat straw are surplus in Punjab, the amount of NPK recycled annually would be about 0.54 Mt. This compares well with the 1.91 Mt/year of fertilizer consumption in Punjab.

Burning of wheat and rice straw has also contributes to loss of soil fertility apart from causing air pollution. For every 4 tons of rice or wheat grain, about 6 tons of straw is produced. Punjab produces around 22 million tonnes of rice straw and 17 million tonnes of wheat straw, annually. The rice crop residue is burnt in the fields instead of its recycling into the soil by mulching it (Jain et al. 2014). This raises the temperature of the soil in the top 3 inches to such a high degree that Carbon-Nitrogen equilibrium in soil changes rapidly. The carbon as CO_2 is lost to atmosphere, while nitrogen is converted to nitrate. Presently, total rice straw utilization in Punjab for different purposes constitutes less than 2% of the 22 million tons produced annually (Yadvinder-Singh et al. 2010). Mishra et al. (2001), while studying the effect of simulated burning condition of crop residues, reported that complete burning of rice straw resulted in 100, 20.1, 19.8, and 80% losses of nitrogen (N), phosphorus (P), potassium (K), and sulphur (S), respectively. The corresponding nutrient losses due to burning of wheat straw were 100, 22.2, 21.8, and 75%, respectively. Besides, nutrient loss, burning of rice straw also results in air pollution. One ton of straw on burning releases 3 kg particulate matter, 60 kg CO, 1460 kg CO_2, 199 kg ash and 2 kg SO_2. The oxide of nitrogen and Sulphur is a potent source of acid rain and its burning produces about 24 million tons of CO_2 in a short span of 15–20 days. Moreover, agriculture experts also maintain that fire in the fields kills friendly pests and bacteria which decrease the fertility of soil.

4.7 Options for Ecosystem Based Management in Agriculture

Ecosystem based management has emerged over the last two decades as a dominant approach to manage natural resources, biodiversity and associated ES. Traditionally sectoral approaches have been applied for the management of natural resources. For example, agriculture land use is managed separately from river bodies, groundwater, forests, urban areas etc. Such approaches often result in conflicts between end users in each sector and are often inadequate in managing natural resources. Therefore, ecosystem based management recognises that all species including animal, plants and human beings are interdependent on each other. It considers that social and economic systems are part of wider ecosystem and optimum interactions between biophysical processes is required for a healthy and functional ecosystem.

Long term sustainability of agriculture in Punjab depends on healthy and functional ecosystems that are managed in an integrated manner. Management of natural resources and biodiversity, which forms the base of current agriculture require changes at policy level to manage landscape as a single ecosystem. There is need to develop an integrated ecosystem based management plan where agriculture sector is not examined individually but a part of wider ecosystem along with forest, water, health sectors. There are several options that are necessary to manage and improve natural resources in the State and in the region so as agriculture sector continues to deliver healthy and nutritional food for all. Some of the policy and practice measures are discussed below.

4.7.1 Crop Diversification

The current rice and wheat rotation system that dominates the agriculture and depletes soil health and groundwater needs to make way for some alternative cropping systems so that the crop diversification can be achieved. Diversification index has decreased significantly over the last 50 years resulting in massive negative impacts on the soil, water and biodiversity. There is growing realisation to change the current rice-wheat rotation. However, more effort is required to provide alternative crops that require less water, help conserve soil and are less susceptible to pests. This requires increased investment in research and development for the improvement of crops, water-use efficiency and nutrient use efficiency.

4.7.2 Biodiversity Policy

Agriculture area dominates the landscape of Punjab. There is very little non-agriculture land available to conserve biodiversity. Moreover, the links between biodiversity in non-farm areas and improvement of agricultural production are clear. For example, pollinators in the margins of farmland and native biodiversity habitats results in better seed set and enhance production of crops. There is need to raise awareness amongst the farming community about the benefits of non-farm biodiversity. A comprehensive state or region wise biodiversity policy that promotes Functional Agricultural Biodiversity (FAB, Wratten et al. 2013) is required to support sustainability in agriculture.

4.7.3 Alternative Agriculture Practices

Global research indicates that there are several alternative agriculture management practices, which are capable of increasing production by protecting natural resources and the environment (Wratten et al. 2013). Some alternative approaches are organic agriculture, conservation agriculture, precision agriculture, agroecology etc. There is need to develop and standardize these practices for Punjab agriculture by working in partnership with farming and scientific community. These interventions will require investments in R&D sector by developing private and public partnerships so that private sector can also contribute towards the sustainable agriculture.

4.7.4 Market Support

There has been much research on the crop diversification side of managing natural resources in agriculture. However, any uptake of new crops by farming community will require market support either by public sector, private sector or by a combination of both.

4.8 Summary

Agroecosystems such as agriculture can provide large amount of ES and can contribute towards food production. However, intensive agriculture such as continuous rice-wheat rotation with high inputs has shown negative impacts on natural resources such as water, biodiversity and soil health as observed in case study of Punjab, India (Shergill 2007). Trade-offs between food production and ES such as soil health, groundwater availability, quality, biodiversity are alarming in this region (Chand 1999; Punjab State Council for Science & Technology 2007). Continuous degradation of natural resources has potential to negatively impact the livelihood of millions of farmers who depend directly on agriculture especially in

Punjab (Rao 2003; Ghuman 2008; Chand et al. 2011). Therefore, a comprehensive policy response is required to safeguard the livelihoods and protection of natural resources so that agroecosystems can continue to supply ES in addition to food production.

References

Abrol IP, Sangar S (2006) Sustaining Indian agriculture—conservation agriculture the way forward. Curr Sci 91:1020–2015

Aggarwal PK, Talukdar KK, Mall RK (2000) Potential yields of rice-wheat system in the indo-gangetic plains of India, rice-wheat consortium paper series 10. Rice-Wheat Consortium for the Indo-Gangetic Plains, New Delhi, India

Bajwa MS (2002) Strategies for agricultural research and development—a plan for Punjab. In: Johl SS, Ray SK (eds) Future of agriculture in Punjab. Centre for Research in Rural and Industrial Development, Chandigarh

Benbi DK, Brar JS (2009) A 25-year record of carbon sequestration and soil properties in intensive agriculture. Agronomy for sustainable development. Springer, Germany

Bhalla GS, Hazell P, Kerr J (1999) Prospects for India's cereal supply and demand to 2020. IFPRI, Washington DC

Bhullar AS, Sidhu RS, Singh M (2006) Integrated land and water use: a case study of Punjab. Econ Polit Wkly 30:5353–5357

Cassman KG, Pingali PL (1995) Extrapolating trends from long-term experiments to farmers fields: the case of irrigated rice systems in Asia. In: Barnett V, Payne R, Steiner R (eds) Agricultural sustainability in economic, environmental and statistical considerations. Wiley, London, UK, pp 63–84

Chand R (1999) Emerging crisis in Punjab agriculture, severity and options for future. Econ Polit Wkly 34:2–10

Chand R, Prasanna L, Singh A (2011) Farm size and productivity: understanding the strengths of smallholders and improving their livelihoods. Econ Polit Wkly 46:5–11

Dale VH, Polasky S (2007) Measures of the effects of agricultural practices on ecosystem services. Ecol Econ 64:286–296

Dhillon BS, Kataria P, Dhillon PK (2010) National food security vis-a-vis sustainability of agriculture in high crop productivity regions.Curr Sci 98:33–36

Directorate of Economics and Statistics (2016) Agricultural statistics at a glance. Department of Agriculture and Cooperation, Ministry of Agriculture, Government of India, New Delhi. http://eands.dacnet.nic.in/

Environment Statistics of Punjab (2011) State's major geographical and environmental features, Government of Punjab

Evenson RE, Gollin D (2003) Assessing the impact of the green revolution, 1960 to 2000. Science 300:758–762

FAO (2011) Save and grow: a policymaker's guide to the sustainable intensification of smallholder crop production. Food and Agriculture Organization of the United Nations, Rome

FAO (2013) The state of food insecurity in the world 2013. The multiple dimensions of food security. Food and Agriculture Organization of the United Nations, Rome

Ghuman RS (2008) Socio-economic crises in rural Punjab. Econ Polit Wkly 43:12–15

Hira GS, Gupta PK, Josan AS (1998) Water logging: causes and remedial measure in south west Punjab. Research Bulletin No. 1/98, Department of Soils, Punjab Agricultural University, Ludhiana

Jain N, Bhatia A, Pathak H (2014) Emission of air pollutants from crop residue burning in India. Aerosol Air Qual Res 14:422–430

Johl SS (1986) Report of the expert committee on diversification of agriculture. Government of Punjab, Chandigarh
Johl SS (2002) Report of the expert committee on agricultural production, pattern, adjustment programme in Punjab for productivity and growth. Government of Punjab, Chandigarh
Johl SS, Ray SK (eds) (2002) Future of agriculture in Punjab. Centre for Research in Rural and Industrial Development, Chandigarh, India
Khan AA, de Jong W, Jansens PJ, Spliethoff H (2009) Biomass combustion in fluidized bed boilers: potential problems and remedies. Fuel Process Technol 90:21–50
Kheti Virasat Mission (2016) http://www.khetivirasatmission.org/
Khurana HS, Phillips SB, Singh B, Alley MM, Dobermann A, Sidhu AS, Singh Y, Peng S (2008) Agronomic and economic evaluation of site-specific nutrient management for irrigated wheat in northwest India. Nutr Cycl Agroecosyst 82:15–31
Kumar P (2002) Economic analysis of total factor productivity of crop sector in the indo-gangetic plain of India by district and region. Agricultural economics, research report, Division of Agricultural Economics, Indian Agricultural Research Institute, New Delhi, India
Mahajan G, Bharaj TS, Timsina J (2009) Yield and water productivity of rice as affected by time of transplanting in Punjab, India. Agr Water Manage 96:(3)
Majumdar K, Jat ML, Pampolino M, Satyanarayana T, Dutta S, Kumar A (2013) Nutrient management in wheat: current scenario, improved strategies and future research needs in India. J Wheat Res 4:1–10
Mathur AS, Das S, Sircar S (2006) Status of agriculture in India: trends and prospects. Econ Polit Wkly 41:5327–5336
MEA (2005) Millennium ecosystem assessment synthesis report. Island Press, Washington
Mishra B, Sharma PK, Bronson KF (2001) Decomposition of rice straw and mineralization of carbon, nitrogen, phosphorus, and potassium in wheat field soil in western Uttar Pradesh. J Indian Soc Soil Sci 49:419–424
Pathak H, Singh R, Bhatia A, Jain N (2006) Recycling of rice straw to improve wheat yield and soil fertility and reduce atmospheric pollution. Paddy Water Environ 4:111–117
Punjab State Council for Science & Technology (2007) State of environment report, Punjab, Chandigarh. http://pscst.gov.in
Rao CHH (2003) Reform agenda for agriculture. Econ Polit Wkly 38:615–620
Rosegrant MW, Cai X, Cline SA (2002) World water and food to 2025: dealing with scarcity. International Food Policy Research Institute, Washington, DC
Sandhu H, Wratten S, Costanza R, Pretty J, Porter J, Reganold J (2015) Global significance of non-traded ecosystem services on farmland. PeerJ 3:e762. doi:10.7717/peerj.762
Sandhu HS, Wratten SD, Porter JR, Costanza R, Pretty J, Reganold J (2016) Mainstreaming ecosystem services into future farming. Solutions 7:40–47
Sanyal SK, Gill MS, Majumdar K (2010) Sitespecific potassium management in rice-based cropping systems in India. In: Proceedings IPI-OUAT-IPNI international symposium on "potassium role and benefits in improving nutrient management for food production, quality and reduced environmental damages", OUAT, Bhubaneswar, 5–7 Nov 2009, pp 389–405
Shergill HS (2007) Sustainability of wheat-rice production in Punjab: a re-examination. Econ Polit Wkly 42:81–85
Sidhu et al. (1998) Changing scenario of Punjab agriculture: an ecological perspective, centre for research in rural and industrial development, Chandigarh
Singh J, Hossain M (2002) Total factor productivity analysis and its components in a high-potential rice-wheat system: a case study of the Indian Punjab. In: Sombilla M, Hossain M, Hardy B (eds) Developments in the Asian rice economy. International Rice Research Institute, Los Baños, Philippines
Singh J, Sidhu RS (2004) Factors in declining crop diversification: case study of Punjab. Econ Polit Wkly 39:5607–5610
Srinivasarao C, Venkateswarlu B, Lal R, Singh AK, Kundu S (2013) Sustainable management of soils of dryland ecosystems of India for enhancing agronomic productivity and sequestering carbon. Adv Agron 121:253–329

Statistical Abstracts (2016) Economic and Statistical Organisation, Government of Punjab, India. http://www.esopb.gov.in/Default.aspx

Swinton SM, Lupi F, Robertson GP, Hamilton SK (2007) Ecosystem services and agriculture: cultivating agricultural ecosystems for diverse benefits. Ecol Econ 64:245–252

Tilman D, Fargione J, Wolff B, D'Antonio C, Dobson A, Howarth R, Schindler D, Schlesinger WH, Simberloff D, Swackhamer D (2001) Forecasting agriculturally driven global environmental change. Science 292:281–284. doi:10.1126/science.1057544

Tilman D, Balzer C, Hill J, Befort BL (2011) Global food demand and the sustainable intensification of agriculture. Proc Natl Acad Sci U S A 108:20260–20264. doi:10.1073/pnas.1116437108

UNEP (2012) Avoiding future famines: strengthening the ecological foundation of food security through sustainable food systems. A UNEP synthesis report, Nairobi

United Nations (2016) United Nations Population Division. http://www.un.org/en/development/desa/population/

Wratten S, Sandhu H, Cullen R, Costanza R (eds) (2013) Ecosystem services in agricultural and urban landscapes. Wiley-Blackwell, Oxford

Yadvinder-Singh, Bijay-Singh, Ladha JK, Khind CS, Khera TS, Bueno CS (2004) Effects of residue decomposition on productivity and soil fertility in rice–wheat rotation. Soil Sci Soc Am J 68:854–864

Yadvinder-Singh, Bijay-Singh, Timsina J (2005) Crop residue management for nutrient cycling and improving soil productivity in rice-based cropping systems in the tropics. Adv Agron 85:269–407

Yadvinder-Singh, Gupta RK, Gurpreet-Singh, Jagmohan-Singh, Sidhu HS, Bijay-Singh (2009) Nitrogen and residue management effects on agronomic productivity and nitrogen use efficiency in rice–wheat system in Indian Punjab. Nutr Cycl Agroecosyst 84:141–154

Yadvinder-Singh, Manpreet-Singh, Sidhu HS, Khanna PK, Kapoor S, Jain AK, Singh AK, Sidhu GK, Avtar-Singh, Chaudhary DP, Minhas PS (2010) Options for effective utilization of crop residues. Directorate of Research, Punjab Agricultural University, Ludhiana, India

Zhang W, Ricketts TH, Kremen C, Carney K, Swinton SM (2007) Ecosystem services and dis-services to agriculture. Ecol Econ 64:253–260

Chapter 5
Desert Ecosystem Management: A Sustainable and Wise Use

Aleem Ahmed Khan, Tahira Ruby, Nargis Naz and Muhammad Rafay

5.1 World Population

The world population depicts an estimated total figure of living humans on Earth 7.136 billion by the United States Census Bureau (UN 2013). The highest rate of growth was observed above 1.8% per year during the 1950s that is peaked at 2.2% in 1963, and then declined to below 1.1% by 2012. Total annual births were highest in the late 1980s at about 138 million, and are now expected to remain essentially constant at level of 134 million, while death rate is 56 million per year, and are expected to increase to 80 million per year by 2040. Current UN projections showed a continued increase in population in the near future that expected to reach between 8.3 and 10.9 billion by 2050 (BBC News 2011; Time 2011). Some analysts have questioned the sustainability of further world population growth, highlighting the growing pressures on the environment, global food supplies, and energy resources (Ozzie 2012).

During the 20th century, the global population saw its greatest increase in known history, rising from about 1.6 billion in 1900 to over 6 billion in 2000. A number of factors are contributing in this increase, including the reduction in mortality rate by improved sanitation, medical advances, and a massive increase in agricultural productivity attributed to the Green Revolution. A number of scientists have argued that the current global population expansion and accompanying increase in resource

A.A. Khan
Department of Zoology, Ghazi University, Dera Ghazi Khan 32200, Pakistan

T. Ruby (✉) · N. Naz
Department of Life Sciences, The Islamia University of Bahawalpur, Bahawalpur 63100, Pakistan
e-mail: tahira.ruby@iub.edu.pk

M. Rafay
Department of Forestry, Range and Wildlife Management, The Islamia University of Bahawalpur, Bahawalpur 63100, Pakistan

H. Sandhu (ed.), *Ecosystem Functions and Management*,
DOI 10.1007/978-3-319-53967-6_5

consumption threatens the world's ecosystem, as well as straining humanity's ability to feed itself (BBC News 2007). The Inter Academy Panel Statement on Population Growth, which was ratified by 58 member national academies in 1994, called the growth in human numbers "unprecedented", and stated that many environmental problems, such as rising levels of atmospheric carbon dioxide, global warming, and pollution, were aggravated by the population expansion. Indeed, some analysts claim that overpopulation's most serious impact is its effect on the environment (UN 2013).

5.2 Indian Sub-continent

There are about 1.7 billion people inhabited in the Indian sub-continent. Geographically, the Indian subcontinent is a peninsular region in south-central Asia, delineated by the Himalayas in north, the Hindu Kush in west, and the Arakanese in east, and extending southward into the Indian Ocean with the Arabian Sea to the southwest and the Bay of Bengal to the southeast. Most of this region rests on the Indian Plate and is isolated from the rest of Asia by mountain barriers. The Indian Plate includes most of South Asia, forming a land mass which extends from the Himalayas into a portion of basin under the Indian Ocean, including parts of South China and Eastern Indonesia, as well as Kuen Lun and Karakoram ranges, and extending up to but not including Ladakh, Kohistan, the Hindu Kush range and Balochistan. Using a more expansive definition—counting India, Pakistan, Bangladesh, Sri Lanka, Nepal, Bhutan and Maldives as the constituent countries—the Indian subcontinent covers about 4.4 million km^2, which is 10% of the Asian continent or 3.3% of the world's land surface area. Overall, it accounts for about 45% of Asia's population (or over 25% of the world's population) and is home to a vast array of peoples (Encyclopedia 2006, 2009).

5.3 Deserts in Indian Sub-continent

Deserts are characterised by the moisture content of the air, which is near zero in most of the deserts, with light rains often evaporated in the dry air before reaching the ground. Rainstorms sometimes come as violent cloudbursts and that may bring as much as 25 cm (10 in.) of rain in a single hour which is the only rain, the desert gets all year (Arshad and Akbar 2002; Akhter and Arshad 2006). The ground heats the air so much that air raises in waves and these shimmering waves confuse the eye, causing travelers to see distorted images called *mirages*. In some deserts, temperatures rise so high that people are at risk of dehydration and even death. At night, these areas cool quickly because they lack the insulation provided by humidity and clouds. Temperatures can drop to 4 °C or lower (Akhter and Arshad 2006).

Many deserts have no drainage to a river, lake, or ocean. Rainwater, including water from flash floods, collects in large depressions called basins. The shallow lakes that form in basins eventually evaporate, leaving playas, or salt-surfaced lake beds. Playas, also called sinks, pans, or salt flats, can be hundreds of kilometers wide. Rain is usually the main source of water in a desert, but it falls very rarely. Many desert dwellers rely on groundwater, stored in aquifers below the surface. Groundwater comes from rain or other precipitation, like snow or hail. It seeps into the ground, where it can remain for thousands of years (Wariss et al. 2013; Akram et al. 1997).

5.3.1 *Thar*

The Thar desert spans an area of approximately 200,000 km^2 and covers large area of both Pakistan and India. It is the seventh largest desert on planet and third largest in Asia. The Thar district is mostly desert and consist of barren tracts of sand dunes covered with thorny bushes (Fig. 5.1). The ridges are irregular and roughly parallel, that they often enclosed sheltered valleys, above which they rise to a height of up to 46 m. These valleys are moist enough to permit cultivation and when not cultivated they yield luxuriant crops of rank grass (Fig. 5.2). But the extraordinary salinity of

Fig. 5.1 Deserts as symbol of life in Thar

Fig. 5.2 Desert as productive rangeland

the subsoil and consequent shortage of potable water, renders many tracts quite uninhabitable (Bakliwal and Wadhawan 2003).

The rain is the main source of water in Tharparkar arid region in Sindh Province, Pakistan. It is very erratic and the annual rainfall varies from 0 to 300 mm. Rainfall in the Thar desert is low—typically between 120 and 240 mm per year—and summer temperatures in July can reach 53 °C (Singhvi and Kar 2004). December, January and February are the coldest months (Cooke et al. 1993). All life activities including agriculture and livestock are dependent on rainfall. The failure of monsoon would mean no agricultural crop and no fodder for the livestock. The main crops of Thar are bajra and guwar. In good rainy years, the lentils, melons and sesame are also grown as mixed crop with the main crop of bajra and guwar. The only hills in the district are at Nagarparkar on the northern edge of the Rann of Kutch which belongs to quite a different geological series. It consists of granite rocks, probably an outlying mass of the crystalline rocks of the Aravalli range. The Aravalli series belongs to Archaen system which constitutes the oldest rocks of the earth crust (Sharma and Kankane 2012).

5.3.1.1 Flora and Fauna

Hydrologically the Thar desert is divided into three main parts viz. humid region, plain lands and integrated stream network (Raman 1997). Approximately 42 families belonged to 82 genera and 125 species of flora were documented in the region. Similarly, 68 species of class Mammalia, 350 species of Aves, 51 species of Reptiles and 08 species of Amphibians were recorded (Sharma and Kankane 2012).

5.3.2 Kharan

The Kharan Desert is located in Northeast of Balochistan. The desert is in the center of a large empty basin also called the 'Sandy Desert'. It covers an area of about 48,051 km^2 and covered with sand dunes, scrub vegetation and weathered rocks (Roberts 2005). The desert consists of moving sand dunes reaching heights from 15 to 30 m with an underlay of pebble floor. The altitude of this desert moves from about 1000 m in the north to approximately 250 m in the southwest. The average rainfall is about 100 mm annually.

The Desert is characterized by very low rainfall, high summer temperature, high velocity winds, poor soils, very sparse vegetation and a low diversity of plant species. The organisms tolerate temperatures as high as 51 °C in summer while winter is generally mild, however, temperature occasionally goes down as low as −10 °C at certain locations (Raziq 2009; IUCN).

The largest dry lake of Balochistan is located in this desert called *Hamun-i-Mashkel*. The 85 km long and 35 km wide area of this lake is filled with sun cracked clay, salthy marshes, moving sand dunes and oxidized pebbles. This part is known for its sudden sand storms and constant mirage (Hagler 2009; IUCN 2000).

5.3.2.1 Flora and Fauna (Kharan Desert)

On the basis of phyto-geography and topographical features, the desert can be distinguished into four habitat types, i.e. mountain/hills, gravel plains, sandy plains and seasonal stream beds. A total of 26 plant species, belonging to 12 families were recorded from the desert (Nawaz et al. 2011). Among faunal diversity, rodents are the most abundant group of mammals along with fox and jackals (Baig et al. 2006). Approximately, sixty species of Herpatile were reported from the area (Khan 1999).

5.3.3 Cholistan

Cholistan desert, locally known as Rohi sprawls covers an area of 16,000 km^2. It adjoins the Thar Desert extending over to Sindh and into India. The dry bed of Hakra River runs through the area, along which many settlements of the Indus Valley Civilisation have been found which is one of the oldest civilizations of the Aryan settlers in the Indian subcontinent. The Cholistan desert extends over an area of 26,000 km^2, in the southern part of the Punjab, Pakistan (Wariss et al. 2013).

On the basis of topography i.e., parent material, soil and vegetation, it is divided into two geomorphic regions. The northern region (Lesser Cholistan, Fig. 5.3) bordering the canal-irrigated areas cover about 7770 km^2 while the southern region (Greater Cholistan) a wind sorted sandy part comprises 18,130 km^2. Vegetation cover is comparatively better in eastern region (200 mm rainfall zone) than the hyper arid southern region (100 mm rainfall zone). The soil topography and chemical composition is playing an important role in plant distribution in the area. The association of certain plant species to certain soils at different places is very common (Wariss et al. 2013; Rafay et al. 2013).

Fig. 5.3 Sandy plains representing moisture conditioning thereby encouring vegetation growth

5.3.3.1 Flora and Fauna

Four types of habitat viz., sand dunes, sandy plains, compact soils with gravels and saline areas are mainly found in the region (Hameed et al. 2002). Regarding floral species, a total of 38 families, 106 genera and 154 species were documented from the whole desert. Among Dicots, 33 families with 79 genera and 115 species are dominated, while monocots include 4 families having 26 genera and 38 species (Wariss et al. 2013). The lesser part is comparatively more diverse dominated by several species of shrubs and perennial grasses while Greater part has poor diversity with sparse vegetation (Akhter and Arshad 2006). Among vertebrate fauna approximately 14 species of class Mammalia, 19 species of Aves and 10 species of Amphibian and Reptiles were reported (Chaudhary et al. 2004).

5.3.4 Thal

The Thal desert is located in northeastern Pakistan between the Indus and Jhelum rivers near the Pothohar Plateau. Its total length from north to south is 190 miles with maximum breadth 70 and minimum 20 miles. The climatic conditions are tropical and has sporadic annual rainfall of 250 mm. Strong gusty winds cause sand storms in summer. Temperatures soar high with summer temperatures going up to 50 °C (Chaudhary et al. 2013).

This region is divided into six districts viz. Bhakkar, Khushab, Mianwali, Jhang, Layyah, and Muzaffargarh. Geographically, it resembles the deserts of Cholistan and Thar. Main towns of Thal are Murshid Abad Shareef, Roda Thai, Mankera, Hyderabad Thai, Dullewala, Mehmood Shaheed, Shah Wala, Shahi Shumali, Piplan, Kundian, Kot Aazam, Sarai Muhajir, Jiasal, Rangpur, Adhi Kot, Jandanaala, Mari Shah Sakhira, Noorpur Thai, Kapahi, Goharwala (Shaheen et al. 2014).

5.3.4.1 Flora and Fauna

The floristic diversity of Thal desert include a total of 38 families with 116 genera and 248 species (Shaheen et al. 2014). The most dominating family was Poaceae, contributed 52 species to the total desert flora including some economically important grass species (Chaudhary et al. 2013). The avian diversity of this desert comprises 55 species and 42 genera belonging to 28 families (Mahboob et al. 2013). The Punjab urial and chinkara are a key wildlife species of the area. Blackbucks were also seen in limited numbers. Four major species viz. one horned rhinoceros, Asiatic cheetah, wild ass, lion, swamp deer and tiger were also observed. Seventeen species of reptiles belonged to thirteen genera were recorded (Khan et al. 1972).

5.3.5 Indus Valley Plains

The Indus Valley Desert is a desert eco-region of northern Pakistan. It covers an area of 19,500 km^2 in northwestern Punjab Province, lying between the Chenab and Indus rivers. The Indus Valley Desert is drier and less hospitable than the Northwestern thorn scrub forests that surround it with temperatures ranging from freezing in winter to extremely hot (more than 45 °C) in summer with only 600–800 mm of rainfall per year. Annual rainfall averages from 640 to 760 mm, slightly more than the Thar Desert.

The mountainous region of valley receives precipitation largely in the form of snow. A large amount of the Indus' water is provided by melting snows and glaciers of the Karakoram, Hindu Kush, and Himalayan mountains. The monsoon rains (July to September) provide the rest of the flow (Fig. 5.4). The climate of the Indus valley ranges from that of the dry semidesert areas of Sindh and lower Punjab to the severe high mountain climate of Kohistān, Hunza, Gilgit, Ladākh, and western Tibet (Sinha 2000).

5.3.5.1 Flora and Fauna

The vegetation is greatly influenced by extreme climatic regime. The desert thorn scrub vegetation is characterized by isolated clumps of *Prosopis* spp., *Salvadora oleoides* and *Caparis* spp., and taller thorn-scrub forests of *Acacia* spp., *Tamarix* spp., *Albizzia lebbek*, and *Morus alba*. This desert ecoregion is not high in richness or endemism, but it does harbor a few large vertebrates that can serve as focal species for conservation. The fauna include wolf *Canis lupus*, hyena *Hyaena hyaena*, caracal *Felis caracal*, leopard *Panthera pardus*, and Punjab urial *Ovis orientalis punjabensis*. The overall mammal fauna consists of thirty-two species, but none are endemic to this eco-region.

Fig. 5.4 A view of vibrant desert profile after monsoon showing clumps of dominating grasses and shrubs

5.4 Functions of Desert

Deserts are important unit of nature and their existence is necessary for mankind and other species. Degradation and deterioration towards deserts threatens quality of life, modern economy and global imbalance in ecosystem. Despite their barren appearance, deserts are teemed with unique and beautiful flora, peculiar fauna which play critical role in healthy functioning of ecosystem, providing goods and services to human life, supporting human economy, supporting life quality and aesthetic lure to human beings.

Deserts are valuable assets as being one of nature's best avenues from the landscape point of view that may include moving sand-dunes, towering bare peaks, plateau-topped rock formations, and smoothly polished ravines. They provide a deep insight into understanding of some ecological phenomenon undergoing in deserts. Local modification of climate and soil around shrubs create and maintain biologically distinct micro-ecosystems. Deserts contribute to national security. The military often uses deserts to test missiles, new airplanes and for training of soldiers. They are the source of outdoor recreation, renewable energy, clean air, water and site station for research.

5.4.1 Deserts Affect Non-desert Climate

It is rather paradoxical that while anthropogenic warming of the global atmosphere already warms and dries deserts, deserts habitually cool the adjacent global atmosphere—a state that is also projected to further intensify, due to global warming. This is due to the desert albedo (the direct reflection of solar radiation by the earth's surface back to outer space). In contrast to the intuition that views the long hours of intense solar radiation reaching the bright desert surface through the dry atmosphere as a cause of enhanced warming, the actual effect of deserts is that of cooling the global atmosphere (Charney 2006). The typical desert albedo is 20–35 per cent of solar radiation reflected back to space, much higher than the 15% of the savannah and the 5% of the rain forest (Verstraete and Pinty 2000).

Global warming is projected to increase desert albedo, through reducing desert vegetation cover, which will further amplify the effect of cooling the non-desert atmosphere and drying adjacent non-desert drylands. The same logic applies also in the opposite direction. "Greening the desert" by restricting grazing or by irrigation would reduce the albedo of the Sahara desert and enhance precipitation over the Sahel; it will also decrease the cooling effect of the desert on the global temperature, thus contributing to anthropogenic global warming (Noble and Gitay 1998).

5.4.2 *Role of Deserts to Mitigate Future Global Warming if Deserts Are Not So Much Dry*

If some desert regions do become significantly moist under global warming, they have the potential to function as a globally significant sink that could tangibly mitigate global warming (Lioubimtseva and Jonathan 2004). On the other hand, those deserts that become drier, with their vegetation only weakly responding to CO_2 enrichment, will not become a significant sink. These deserts are also not likely to act as a significant source driven by land degradation, because the turnover rate of the large desert sink of inorganic soil carbon is too slow to generate significant CO_2 emissions. Also, although the turnover of soil organic carbon is fast and land degradation in deserts might increase CO_2 emissions (as the carbon in eroded soil is oxidized), the pool of soil organic carbon that might be affected by land degradation is too small to make this a significant contribution to global atmospheric CO_2. Between-ecosystem comparison, the scenario of further desert warming and reduced precipitation, the ratio of soil organic carbon to soil inorganic carbon in deserts will be reduced. Soils of deserts sequester carbon in inorganic and organic compounds. It is the organic carbon that is most readily sequestered—rainfall allowing—with rates of accumulation of 5–10 g cm^{-2} $year^{-1}$ under best-practice rain-fed farming in arid-semiarid regions (Lal 2002).

5.5 Management of Deserts

5.5.1 *Management of Surface and Groundwater Resources in Cholistan Desert of Pakistan for Drinking and Farm Production*

The main sources of water in Cholistan desert are rain and groundwater. The rainwater is collected in tobas (ponds) and Kunds for drinking and domestic use for a maximum of 3–4 months. The ground water is obtained through dug wells and tubewells. Therefore, people are used to migrate towards canal irrigated areas alongwith their livestock till next rainy season when water from small ponds is exhausted (Akram et al. 1997).

5.5.2 *Range Management as a Discipline in Desert*

The principles of range management cover the entire spectrum of Range Science i.e. philosophy, site physiology, soil science, biology of range plants and livestock (Qureshi et al. 1993). They include:

- Principles related to Man
- Principles related to Range site
- Principles related to Range vegetation
- Principles related to Range livestock.

Objectives of Range management include:

Ancestral profession, sacred, memorial, way of life, status symbol, recreation, sanctuary, landscaping, soil and water conservation, research, demonstration, political motives and commercial enterprise.

5.5.3 Thal Desert Range Area Management

It includes planned grazing, deffered rotational, close use with long rest period, grazing interval not less than half month, two years protection to reseeded area, providing limited irrigation on selected favourable sites where possible, reseeding in favourable sites, hay-making on reseeded areas and on irrigated areas, planting of fodder trees on selected sites, stock water development and digging of shallow wells (Qureshi et al. 1993).

5.5.4 Thar Desert Range Area Management

It includes grazing according to carrying capacity, mixing sand in clayey flat by their ploughing across summer winds, provide limited irrigation in selected favourable sites, planting of fodder trees, stock water development: shallow wells, ponds, ponds with adjacent wells, windmills etc. (Qureshi et al. 1993; Rafay 2012).

5.5.5 Cholistan Desert Range Area Management

It includes grazing at suitable stocking rate, artificial reseeding of grasses on selected favourable sites, hay-making, cropland for fodder production, forest lands to be opened for grazing, planting of fodder trees, water conservation on selected favourable sites and stock water development (Qureshi et al. 1993).

Fig. 5.5 An overview of nomadic life in desert

5.5.6 Dry Land Environmental Management of Balochistan Deserts 'Kharan and Chagai'

The major issues of the district are: loss of vegetative covers in the watersheds and rangelands due to overgrazing, and use of wood for fuel by the local and nomadic population (Fig. 5.5) is the major concern of this area (IUCN 2000). Soil loss and reduction in fertility due to wind and water erosion and limited water availability and its poor quality for domestic, agricultural and other requirements also effect the economy of this area.

Few general management approaches include: sustainable land use planning and management practices by village and pastoral communities and development of national and regional policies.

The specific recommendations include: establishing the impacts of climate variability and change on the fragile economy of dryland households and farming systems, establishing pilot areas to help conduct case studies for identifying the typology of primary measures undertaken at household and community level to adapt to variable and changing climate, working with selected households and partners to evaluate the potential for rainwater harvesting and conservation technologies as an adaptive measure to water scarcity, demonstrating and characterizing effective rainwater harvesting methods for crops and pastures.

5.6 Desertification Control for Sustainable Land Use

Desertification, a process of gradual land degradation afflicting many arid, semi-arid, and dry sub-humid areas around the world, can lead to environmental deterioration, natural resources depletion, agricultural losses or food shortages and substantial hardships for affected populations (Hare 1977). Deserts all over the world are known to be increasing at the expense of agricultural land and this is most severe in arid and semi-arid regions. The desertification leads to agricultural productivity losses and increases poverty. It also causes significant reductions in carbon storage in soils, contributing to global warming, and loss of biodiversity. It triggers soil erosion because of the loss of vegetative ground cover exacerbating water erosion and flash floods, accelerating siltation of rivers and reservoirs (Kassas 1977).

In developing world, the key driving force behind desertification is a nexus of poverty, rapid population growth, and inadequate progress in increasing agricultural productivity. With slow agricultural and economic growth, the pressure on natural resources in dryland areas seems destined to worsen. Given either no or low and erratic rainfall, the scope for absorbing all these additional workers in agriculture whilst maintaining or increasing per capita incomes and food supplies will be difficult in years to come and is going to be a dream to be realized (Rozanov 1981). A profound change in local and international behaviour should be made to address the livelihood needs of dryland populations and reversal of desertification process on war footings through short term goals and long term initiatives with a political will and commitment.

5.6.1 Range Livestock Improvement

It includes few expensive operations which increase the system potential and help overcome any unexpected catastrophe e.g. improve quality of herd by artificial insemination, importing better quality animals, arrangements of emergency feed during periods of femines, use of supplementary feed to accelerate the growth rate of animals, arranging necessary veterinary aid for protection against epidemics and arranging transport of animals out of the drought stricken range to market for quick disposal.

5.6.2 Range Vegetation (Vegetation Is a Basic Resource Which Supports Livestock) Improvement

It includes protection of vegetation against grazing, browsing, cutting lopping etc. is done by erecting fences, constructing stone walls, establishing live hedges and support of law enforcing authorities. Selection of suitable livestock species by keeping in mind the grazing behavior, present and desired vegetation composition, slope of the area, type of soil, climate and market etc. Suitable grazing seasons are determined by keeping in view the growth cycles of major range forage species and avoiding grazing during extreme climate conditions. Seasons of grazing with frequency and intensity of grazing can thus be manipulated to maintain a desired floristic composition of range vegetation. Season of grazing has strong influence on nutritive value, palatability and digestibility of forage species (Qureshi et al. 1993). In addition artificial reseeding of local high yielding and palatable grasses for summer and winter, artificial reseeding of legumes, water spreading, water conservation and fertilization of extensive range areas. Intensive forage production with irrigation and fertilization in a restricted cultivated area is promoted. Breeding techniques are introduced to develop new forage crop varieties.

5.6.3 Pastoralism

Along with range resources, the agricultural farms also utilized by livestock but the importance of crops varies according to the status of small owner-pastoralists. After the monsoon, when sufficient forage is available, the livestock move to desert rangelands for grazing (Akhter and Arshad 2006). Their livestock comprising mainly buffaloes, cattle, some sheep and goats restricted to the irrigated fields where feed resources comprise a variable proportion of crop residues, fodders and peri-irrigation grazing areas. The seasonal movement of livestock and people to the desert and subsequently again towards irrigated margins was frequently observed phenomenon. With the onset of monsoons in July/August when the *tobas* (Fig. 5.6) are filled with water, the nomads start moving towards desert. The animals move with the transhumant livestock from one *toba* to the next and subsequently to the wells and/or *kunds*. However, they remain at the wells or *kunds* when the transhumant livestock retreat to the irrigated margins.

Fig. 5.6 Post-monsoon desert water reservoir, locally called Toba

5.7 Concluding Remarks

Key deserts explored in Indian sub-continent performs several ecological functions and play an important role to support livelihoods of millions of its inhabitants. Deserts are fragile ecosystems that are vulnerable to climate change. Any change in climate will be further detrimental to the vast biodiversity of deserts. Therefore, there is need to implement ecosystem based management to protect desert biodiversity and other natural resources. This can be achieved by sustainable use of deserts for human activities by implementing strategies to harvest and use water more efficiently, sustainable crop production, grazing for livestock and integrated desert ecosystem management.

References

Akhter R, Arshad M (2006) Arid rangelands in Cholistan desert (Pakistan). Scheresse 17:210–217

Akram MM, Abdullah W, Khan A, Sheikh BA (1997) Management strategies for water resources development in Cholistan desert for human, livestock and small scale agriculture. In: International symposium on water for the 21st century, Lahore, June 17–19, pp 121–130

Arshad M, Akbar G (2002) Benchmark of plant communities of Cholistan desert. Pak J Biol Sci 5:1110–1113

Baig KJ, Awan MR, Ashraf N (2006) Ecological studies and zoogeographic affinities of the amphibians and reptiles found in Chagai desert, Balochistan, Pakistan. Pak J Zool 38:145–151

Bakliwal SK, Wadhawan SK (2003) Geological evolution of Thar desert in India -issues and prospects. Proc Indian Nat Sci Acad 69:151–165

BBC News (2007) The limits of a green revolution? Mar 29

BBC News (2011) Population seven billion: UN sets out challenges

Charney JG (2006) Dynamics of deserts and drought in the Sahel. Q J R Meteorol Soc 101: 193–202

Chaudhary SN, Arshad M, Ahmad E, Mustafa G, Fatima S, Akhtar S, Amjad MS (2013) Ethno-botanical evaluation of grasses from Thal desert, Pakistan. Sciences 66:1661–1664

Chaudhary MS, Sial N, Naseema A (2004) Natural resources and their utilization, with special reference to Cholistan desert, Pakistan. Bio-Sci Agric 9:1–4

Cooke R, Warren A, Goudie A (1993) Desert geomorphology. UCL Press, London

Encyclopedia (2006) Indian subcontinent. Encyclopedia of modern Asia (2006)

Encyclopedia (2009) "Asia" overview. In: Encyclopaedia Britannica (2009) Encyclopaedia Britannica Online

Hagler B (2009) The biophysical survey of Tahlab and Hamun-e-Mashkhel Wetlands.Unpublished Report. p 5

Hameed M, Chaudhary AA, Maan MA, Gill AH (2002) Diversity of plant species in Lal Suhanra National Park, Bahawalpur. Pak J Biol Sci 2:267–274

Hare K (1977) Connections between climate and desertification. Environ Conserv 4:82

IUCN Pakistan (2000) Balochistan partnerships for sustainable development Balochistan Programme Office Marker Cottage. Zarghoon Road Quetta, Pakistan

Kassas M (1977) Arid and semi-arid lands: problems and prospects. Agro-Ecosystems 3:186

Khan NU, Ilyas M, Rahman W, Mashima T, Okigawa M, Kawano N (1972) The flavonoids and their role in different diets. Tetrahedron 28:5689–5690

Khan MS (1999) Herpetology of habitat types of Pakistan. Pak J Zool 31:275–289

Lal R (ed) (2002) Encyclopedia of soil science. Marcel Dekker, New York, p 1476

Lioubimtseva E, Jonathan A (2004) Possible implications of increased carbon dioxide levels and climate change for desert ecosystems. Environ Manag 33:388–404
Mahboob S, Nisa Z, Balawi A, Minsed F, Ahmad Z, Sultana S (2013) Study on Avian diversity of Thal desert, Punjab, Pakistan. Life Sciences J 10(11)
Nawaz A, Rafique M, Khan NK (2011) Pattern of mammalian distribution in the Chagai desert, Balochistan, Pakistan. Pak J Zool 43:841–847
Noble H, Gitay I (1998) Climate change: impact, adaptation and vulnerability. McGraw Hill Publications, USA
Ozzie Z (2012) Green illusions. University of Nebraska Press, Lincoln and London, pp 187–331
Qureshi MA, Khan G, Yaqoob M (1993) Range management in Pakistan. Kazi publications, Lahore, Pakistan, p 229
Rafay M (2012) Studies on productive potential and conservation strategy of major range grasses in the degrading rangelands of Cholistan Desert. Ph.D. thesis, University of Agriculture, Faisalabad, Pakistan
Rafay M, Khan RA, Yaqoob S, Ahmad M (2013) Floristic composition of grass species in the degrading rangelands of cholistan desert. Pak J Agric Sci 50:599–603
Raman CV (1997) Thar desert ecosystem. Curr Sci 72:1
Raziq A (2009) Assessing the potential of the indigenous livestock breeds of Balochistan. Drynet: a science and technology expertise. Project study report, funded by the European Union and supported by The Global Mechanism
Roberts TJ (2005) Field guide to the small mammals of Pakistan. Oxford University Press, Oxford, p 250
Rozanov BG (1981) Degradation of arid lands in the world and international cooperation in desertification control. Pchvovedenie 8:5–11
Shaheen H, Qureshi R, Akram A, Gulfraz M, Potter D (2014) A preliminary floristic checklist of Thal desert Punjab, Pakistan. Pak J Bot 46:13–18
Sharma G, Kankane PL (2012) Fauna of ecosystems of India, Thar desert. Special publication series on the occasion of CBD Cop-11, India
Singhvi AK, Kar A (2004) The aeolian sedimentation record of the Thar Desert. Proc Indian Acad Sci 113:371–401
Sinha A (2000) The ancient Indus Valley civilization. A guide to accompany Indus River: journey of a Lifetime. WGTE Public Broadcasting, 1270 S Detroit Avenue, Toledo
Verstraete MM, Pinty B (2000) In: Environmental information extraction from satellite remote sensing data, vol 114. American Geophysical Union, Washington, D.C., p 7
Wariss HM, Mukhtar M, Anjum S, Bhatti GR, Pirzada SA, Alam K (2013) Floristic composition of the plants of the Cholistan Desert, Pakistan. Am J Plant Sci 4:58–65
TIME (2011) Overpopulation's real victim will be the environment
United Nations (2013) Department of economic and social affairs, population division. United Nations, New York, USA

Chapter 6
Forest Ecosystem: Functions, Value and Management

Madhu Verma, Rohit Singh and Dhaval Negandhi

6.1 The Status of Forests in the South Asia

South Asia is a land of diversity. A 4.7 million square kilometer strong region with a GPD of $2.35 trillion and a population count of 1.7 billion individuals in 2013 (World Bank 2014), South Asia is a region not to be missed in any ways. A few kilometers of travelling might change everything from culture to languages to economics to ecology. Probably this is the reason why South Asian countries attract so many tourists, over 10 million international tourists in 2012 (World Bank 2014), from round the globe that come to explore the variations. Another notable feature about the countries of this region is rapid social and economic change. The dilemma between conservation and/or development is arguably more pronounced here that in most parts of the globe. In the world where the already developed nations are pushing for conservation, this developing region is struggling to meet out its development imperatives. Development having its own implications on the ecology, the more these countries develop, the more likelihood of negative impacts it would have on the ecology of the region.

South Asia covers about 10 per cent land area of the Asian continent but accounts for more than 40 per cent of its population. The figure suggests the extent of population pressure per unit area and scarcity of resources among the population. Agriculture is the major land use i.e., 2.6 million square kilometers (World Bank 2014) in this region and accounts for more than half of the total land use. Climatic differences, dependency on rain for irrigation and degrading soil conditions are among the major limiting factors to the agricultural productivity. Uneven and erratic rainfall patterns in recent past have further exacerbated the problem. Forests

M. Verma (✉) · R. Singh · D. Negandhi
Indian Institute of Forest Management, Bhopal, Madhya Pradesh, India
e-mail: mverma@iifm.ac.in

H. Sandhu (ed.), *Ecosystem Functions and Management*,
DOI 10.1007/978-3-319-53967-6_6

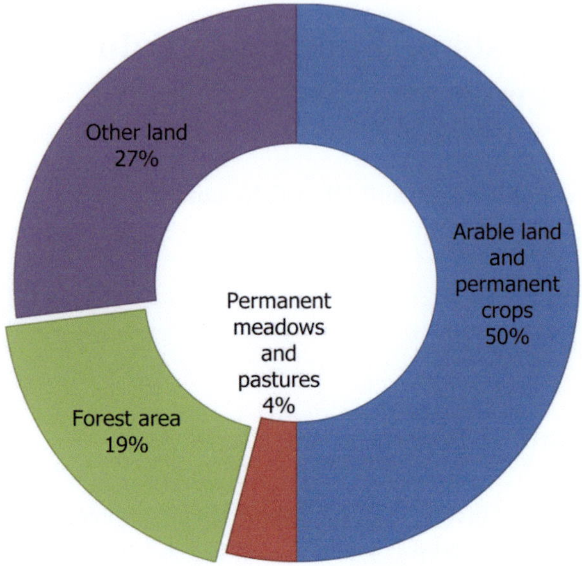

Fig. 6.1 Land use of South Asia. *Source* Modified FAO (2012)

Table 6.1 Main forest types in South Asia

Country/Type of forests	Bangladesh	Bhutan	India	Nepal	Pakistan	Maldives	Sri Lanka
Tropical wet evergreen forests	■		■				■
Tropical semi evergreen forests	■		■				■
Tropical moist deciduous forests	■		■	■			■
Tropical dry deciduous forests			■		■		■
Tropical thorn forest			■		■		■
Littoral and Swamp forests	■		■		■	■	■
Subtropical broadleaved hill forests	■	■	■	■	■		
Subtropical pine forests	■	■	■	■	■		
Subtropical dry evergreen forests			■		■		
Montane wet temperate forests			■				■
Himalayan moist temperate forests		■	■	■	■		
Himalayan dry temperate forests		■	■	■	■		
Alpine and Subalpine forests		■	■	■	■		

Source Various APFSOS II country papers

occupying 0.8 million square kilometers (FAO 2012) accounts for 19 per cent of the land use in South Asia, next only to agriculture. The land use pattern of the region is broadly depicted in Fig. 6.1.

The forests of South Asia are no different to variations. The diversity among forests can be judged from the fact that nearly all major types of forests can be found in this region (Table 6.1).

Forests in this region not only have ecological significance, but also critical socio-economic functions. They provide various economic goods and resources in

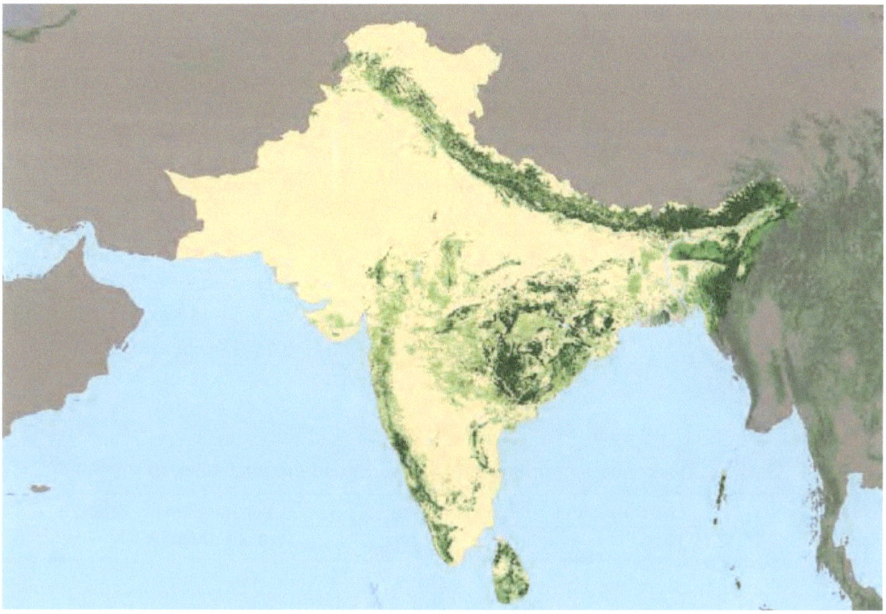

Fig. 6.2 Forest cover in South Asia 2005. *Source* FAO (2012)

Table 6.2 Extent of Forest in South Asia in 2010

Country	Forest area ('000 hectare)	Forest area (% of geographical area)	Forest area (hectare) per 1000 persons
Bangladesh	1442	11	9
Bhutan	3249	69	4750
India	68,434	23	58
Maldives	1	3	3
Nepal	3636	25	126
Pakistan	1687	2	10
Sri Lanka	1860	29	93
South Asia	80,309	19	51

Source FAO (2012)

addition to ensuring flow of intangible benefits through ecosystem services such as timber, fuel and fiber, clean air, climate control, water purification, and carbon sequestration among several others. The forests are also livelihood sources for a large number of communities that depend upon them. Considering the high socioeconomic dependence of people in this region, there has been considerable decentralization in the management of forests in South Asia, albeit, significant national variations do exist. The map shows the forest cover in South Asia (Fig. 6.2).

Table 6.3 Changes in forest area over time

Country/Region	1990–2000		2000–2005		2005–2010	
	ha ('000)/year	%	ha ('000)/year	%	ha ('000)/year	%
Bangladesh	−3	−0.18	−3	−0.18	−3	−0.18
Bhutan	11	0.34	11	0.34	11	0.34
India	145	0.22	464	0.70	145	0.21
Maldives	0	0.0	0	0.0	0	0.0
Nepal	−92	−2.09	−53	−1.39	0	0.0
Pakistan	−41	−1.76	−43	−2.11	−43	−2.37
Sri Lanka	−27	−1.20	30	−1.47	−15	−0.77
South Asia	−7	−0.01	347	0.44	96	0.12

Source FAO (2012)

Table 6.4 Growing stock volume (million m^3 over bark) and growing stock in (m^3)

Country	Growing stock in forests per ha			Per ha growing stock in 2010 (m^3)
	1990	2000	2005	
Bangladesh	72	71	70	48
Bhutan	535	592	650	200
India	4363	4662	5489	80
Nepal	856	694	647	178
Pakistan	261	211	160	95
Sri Lanka	57	47	39	21

Source FAO (2012)

India, being the largest nation on land area, accounts for more than 85% of the total forests of South Asia. However, when it comes to the total percentage of geographical area under forests, India stands forth behind Bhutan, Sri Lanka and Nepal (Table 6.2). In terms of highest per capita forest in South Asia, Bhutan has close to 5 ha of forest per capita as compared to Maldives which has about 3 ha of forest per 1000 persons. Table 6.3 shows the changes in the forest area over the years. This is an interesting gradient as it reflects how the countries have fared over the dilemma of conservation versus development in the region.

But, the change in the forest area is not the only parameter to judge the efforts of the countries on conservation. The lack of good management practices, increasing human pressure from fuel wood collection and grazing added up by fires, pests and diseases may result in degradation of the quality of forests. The change in growing stock in these countries over time can be seen in Table 6.4. On the ownership of forest lands, it can be seen that the majority share of the forests in this region is owned by the Government (Fig. 6.3). India which has the majority of land under forest in South Asia has 86 per cent forests owned by the Government.

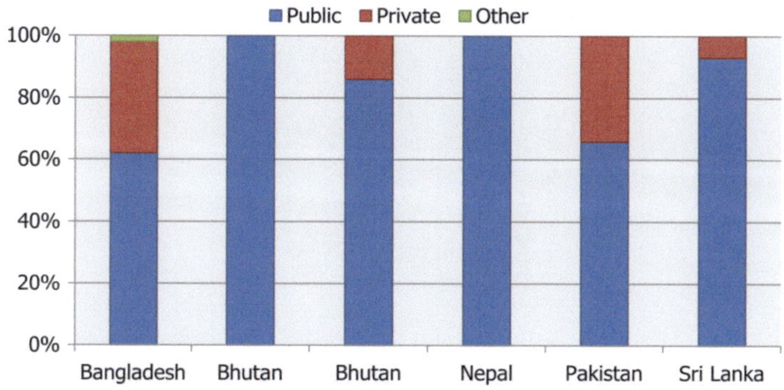

Fig. 6.3 Ownership of forests in South Asia. *Source* FAO (2012)

6.2 Important Functions (Services) of Forests

Forests are critical resources of a country in terms of their social, environmental and economical benefits (Table 6.5). They not only provide a variety of goods such as fuel wood, timber, pulpwood, fodder, non-wood forest produce and act as basic sources of raw materials for industries and other commercial activities but also provide an array of ecosystem services which are pre-requisite for ensuring human well-being (and the sustenance of other life forms) on this planet (Fig. 6.4). The benefits of these ecosystem services such as provisioning of clean air, recharge of groundwater and its purification, carbon sequestration and many more are not limited to the area under forest cover but often extend well beyond. Depending on the ecosystem service, the service shed of these services may be regional, national or even global.

Following the Millennium Ecosystem Assessment framework (MEA 2005), the ecosystem services from forests may be categorized into the following types:

6.2.1 Provisioning Services

Forests of South Asia bear a lot of products which are mostly harvested by the forest communities as part of their livelihood activities (see case study of mangroves: a livelihood option for the coastal communities of Bangladesh in Sect. 6.3). The food products include fruits and vegetables found inside the forests. The collection of non-wood forest produce is a practice which includes products such as honey and silk. Collection of fuel wood is a common practice, which is used for cooking and other purposes by the households dependent on the forests. Other than these, South Asia's biodiversity richness means that a lot of medicinal plants exits in the forests which are used as biochemical, pharmaceuticals and natural medicines (see case study on bioprospecting and conservation of genetic resources in

Table 6.5 State of world's forest (FAO 2012)

Country/Area	Employment		Gross Value Added		Food Security		Energy		Shelter	
	Total for the formal and informal sector		Total for the formal and informal sector		Number of people using wood fuel to cook		Primary energy supply from wood		People in homes made partly from forest products	
South Asia	('000)	(% of total labour force)	(US$ million)	(% of total GDP)	(million)	(% of total population)	(MTOE)	(% of TPES)	(million)	(% of total population
Bangldesh	734	1	2011	1.7	59	39	7	23	0.3	0.2
Bhutan	318	85	273	15.6	0.2	36	1	92	0.1	15
India	4751	1	36,511	2.0	625	50	96	13	191	15
Maldives	–	–	–	–	0.04	14	0	1	0.1	45
Nepal	125	1	212	1.2	19	64	3	31	8	25
Pakistan	380	0.6	1560	0.8	101	57	12	14	59	33
Sri Lanka	47	0.5	1298	2.2	14	69	1	13	–	–
	6355	1	41,864	1.9	820	51	121	14	259	16

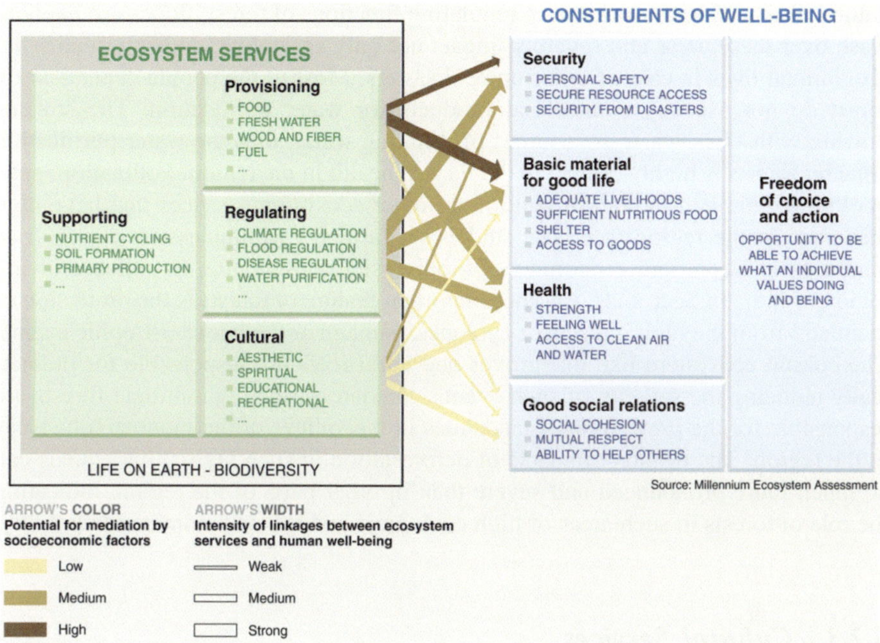

Fig. 6.4 Ecosystem services and the constituents of human well-being. *Source* MEA (2005)

Sect. 6.3). Forests are also responsible for water provisioning services where they act as watersheds. Another service provided by forest is the water purification which is crucial for the people living in vicinity and using streams originating from forests for drinking water.

6.2.2 Regulating Services

Forests ensure flow of a range of intangible regulating services. They help maintaining the air quality by allowing deposition of affluents from the air. The regulations on industrial emissions are less stringent in this region and forests play a major role in maintaining a healthy breathing environment for the people saving them from respiratory and other disorders. Forests also help in the regulation of a number of diseases in the human population by controlling the abundance of disease vectors and human pathogens. Besides these, forests play a vital role in climate regulation by sequestering carbon. Water regulation is an important aspect looking at the demographics of the region. Forests mitigate flooding in downstream inhabited areas (see case study of Uttarakhand disaster in Sect. 6.3). Within South Asia, Bangladesh is the major sufferer of floods followed by India and Pakistan. The forests in these countries have however helped in abating the magnitude of these natural disasters. In the absence of regulating functions of forest, floods are likely to wash over these areas and severely impact not only agriculture and soil quality but also human lives in case of unfortunate disasters. Most of the people dependent on forest do not use any commercial products for water purification. The streams flowing with fresh water are sources of drinking water, thus the water purification done by forests is highly critical to the human health in the region. Pollination, gene pool protection and biological control are other ecosystem services that have high relevance in the region (see case studies on farmer's willingness to pay for rice landraces in Nepal and economics of biodiversity conservation in coffee growing region of India in Sect. 6.3). Another important feature of forests is the protection of mainland from the high impacts of cyclones, tsunami and other catastrophic events. The coastal ecosystem like mangroves and coral reefs are responsible for dramatically reducing the severity of such events. In contrast, the high-altitude forests are responsible for the prevention of landslides in the valleys of the mountainous areas in the region. The negative impacts of deforestation in such high-altitude areas can be much more pronounced and severe than in other parts of the region, indicating the role of forests in such areas of high ecological and socioeconomic vulnerability.

6.2.3 Cultural Services

South Asia is bestowed with rich cultural heritage and forests are an intricate part of the culture in the region. Be it the spiritual aspect, recreation, heritage or the aesthetic values people derive from it, a range of cultural services emanate from forests in South Asia (see case study of relationship between environmental status and gross national happiness index in Bhutan in Sect. 6.3). Sacred groves and sacred hills are a notable feature of this region. The religious institutions inside the forest areas have high spiritual values to not only the people of forest community but for people throughout a country or the region and hence such places attract high pilgrimage throughout the year. Some of such institution are centuries old which also have high heritage values (see case study on biodiversity values, status and strategies for the conservation of sacred groves in Indian in Sect. 6.3).

Recreation is another ecosystem service that forests provide. People are now willing to spend more on the leisure activities in such forest areas (see case study of Sukkomajri—a case of forest and watershed protection in India in Sect. 6.3). Considering biodiversity richness in terms of both flora and fauna, tourism in predominantly forest areas is thriving in the region. For example, to get a glimpse of tigers of India and Bangladesh, the tourist are attracted not just from South Asia but from across the globe. Forest also offers opportunities to preserve and use indigenous knowledge that has been an important part of the communities inside the forest areas and in vicinity. The importance of indigenous knowledge for medicinal plants, adaptation to climate change, biomimicry and other fields is being increasingly recognized by the modern day science.

6.2.4 *Supporting Services*

Apart from provisioning, regulating and cultural services, forests ensure basic functions such as soil formation, nutrient cycling and primary production that are fundamental for the flow of other ecosystem services. The production of atmospheric oxygen, nutrient composition of the soil, provisioning of habitats are some of the vital ecosystem services from forests that support humankind as well as other life forms on this planet.

6.3 Influential Cases of Ecosystem Services and Human Wellbeing Linkage

6.3.1 *Uttarakhand Disaster*

Uttarakhand Disaster is an example of how conversion/diversion of mountain ecosystem can lead to the extreme and catastrophic events. The rapid expansion of hydro-power projects, ever increasing tourism, mainly religious, which has promoted the expansion of road network and establishment of unplanned hotels, shops and multistory housings in ecologically sensitive zones are primarily the human activities which propelled the magnitude of disaster. About 5000 people died in Himachal Pradesh and Uttarakhand, India due to floods in 2013. Ministry of Environment later issued a statement that the National Ganga River Basin Authority had notified 130 km stretch from Gaumukh to Uttarkashi as an ecosensitive zone on December 18, 2012. The notification, thus, prohibited activities such as setting up of hydroelectric power plants of more than 25 MW, extraction of river water for new industrial purposes, mining except for domestic needs, stone quarrying, deforestation, burning of solid waste.

6.3.2 *Bioprospecting and the Conservation of Genetic Resources*

Bioprospecting has been touted as a source of finance for biodiversity conservation (Rausser and Small 2000). The bioprospecting value of the "marginal unit" of genetic resources is likely to be vanishingly small, creating essentially no conservation incentive. Scientific models channel research effort toward leads for which the expected productivity of discoveries is highest. Leads of unusual promise then command information rents, associated with their role in reducing the costs of search. When genetic materials are abundant, information rents are virtually unaffected by increases in the profitability of product discovery and decline as technology improvements lower search costs. Under plausible conditions, the

Table 6.6 Bioprospecting values in several ecosystems in south Asia as a function of density of endemic species (Rausser and Small 2000)

Biodiversity "Hot Spots"	Forest area (1000 ha)	Density, endemic species/1000 ha	Hit probability (/1000 ha)	Incremental value ($/ha)	Simpson et al. scarcity rent ($/ha)
Southwestern Sri Lanka	70	7.14	8.57E^04	7463	16.84
Western Ghats of India	800	2.03	2.44E^05	2026	4.77
Eastern Himalayas	5300	0.42	5.04E^06	332	0.98

bioprospecting value of certain genetic resources could be large enough to support market-based conservation of biodiversity (Rausser and Small 2000). Table 6.6 summarises bioprospecting values in several ecosystems in south Asia.

6.3.3 Household Characteristics and Forest Dependency in Nepal

In resource dependent rural areas of developing countries, common property resource management has been considered as one of the most viable options for combining poverty reduction, enhancement of local level economic development and biodiversity conservation (Adhikari and Lovett 2004). The collection and extraction activities in common forests (forests that are used by local communities living in or in vicinity of the forests for their sustenance) are correlated with household socio-economic attributes. But equity of resource distribution is important parameter for the overall development. Households belonging to higher caste derive more resources than the lower caste. Appropriate cost-benefit sharing arrangements, together with empowerment of resource users, technical assistance to develop and strengthen local organizational capacities, and support equitable and sustainable management efforts are examples of other essential elements (Adhikari and Lovett 2004). Empowerment of resource users refers to mechanisms that allow direct involvement of vulnerable and marginalized groups in common forests activities in terms of representation of these groups in the decision making authority as well as representation of their interest in forest management regimes.

6.3.4 The Economics of Biodiversity Conservation in Coffee Growing Region of India

In the context of a tropical ecosystem in India, where coffee is the one of the main competitors for land use, the opportunity costs of biodiversity conservation in terms of coffee benefits foregone can be quite high (Ninan and Sathyapalan 2003). In a study conducted in a coffee growing region of India, it was found that even after including external costs due to wildlife damages and defensive expenditures to protect against wildlife, the Net Present Value and Internal Rate of Return from coffee for all land holding groups were quite high (Ninan and Sathyapalan 2003). The external costs accounted for between 7 and 15% of the total discounted costs of coffee cultivation, and smaller holdings proportionately incurred higher external costs as compared to larger holdings. High transaction costs incurred by the growers to claim compensation for wildlife damages and thus the local communities were also willing to pay in terms of time for participatory biodiversity conservation but preferred a decentralized government institution for the purpose.

6.3.5 Farmers' Willingness to Pay for Rice Landraces Conservation in Nepal

Crop genetic resources constitute an important aspect of biodiversity conservation, both because of their direct value to the farmers and due to their indirect global value. Contingent valuation has been used to document the economic value of crop genetic resources based on the farmers' willingness to pay for conservation. The average mean willingness to pay was USD 4.18 for in situ and USD 2.20 for ex situ conservation per annum (Poudel and Johnson 2009). Landholding size, household size, education level, socio-economic status, sex of respondent, number of crop landraces grown, and knowledge on biodiversity influences the willingness to pay for in situ conservation, whereas only landholding size and household size influences the willingness to pay for ex situ conservation. Economic valuation of crop genetic resources can be seen as a tool to provide assistance to the policy makers in setting conservation priorities (Poudel and Johnson 2009).

6.3.6 Mangrove: A Livelihood Option for the Coastal Community of Bangladesh

Mangrove ecosystem provides both goods and services for coastal community, helps to improve livelihood options and protect them from natural disaster by providing variety of environmental support (Sarker 2010). Besides livelihood, mangrove forests play an important role in maintaining healthy coastal environment. It serves as a protection for a myriad of juvenile aquatic species, functioning as a habitat for a variety of terrestrial fauna and a source of nutrients that helps to sustain many complex food chains. Up to 80% of the fish catch is directly or indirectly dependent on mangroves of Sundarbans (Sarker 2010).

6.3.7 *Biodiversity Value, Status and Strategies for the Conservation of Sacred Groves in India*

The people of Manipur, a state in northeast India, follow ancestral worship and animism in the form of deity worship, with the central focus on worship in forest patches. The beliefs and taboos associated with the *Sylvan* deities (*Umanglais*) in the forest patches are restricted to any sort of disturbance of flora and fauna (Khumbongmayum et al. 2005). These social boundaries help to conserve the entire organism as a whole, which stand the concept of sacred groves. Well-preserved sacred groves are store houses of valuable medicinal and other plants having high economic value, and serve as a refuge to threatened species. 96% of the species existent are being used as medicine for the treatment of various ailments. A few of the medicinal plants which have disappeared from the locality are now confined only to the groves. Therefore, conservation measures of sacred groves need to be formulated considering the factor of degradation and the basic necessities of the local people. Until and unless a viable option is provided to the local people (especially those who habitat nearby the adjoining areas) for sustaining their economic condition, no step for conservation of biodiversity will be successful (Khumbongmayum et al. 2005).

6.3.8 *Relation Between Environmental Status and Gross National Happiness in Bhutan*

The Himalayan Kingdom of Bhutan is in the midst of transformation as it moves from an isolated past to a modern nation-state and participant in the global community. Its development path embraces the concept of "Gross National Happiness," a philosophy and policy instrument that seeks to promote human development and manage environmental conservation within a sustainable strategy guided by Buddhist ethics (Zurick 2006). Several concerns challenge the model of GNH. The goal of cultural preservation requires the resolution of ethnic conflicts, including those involving the Nepali residents and refugees. Economic development based on natural resource extraction will require effective enforcement of regulations embedded in the conservation policies; such enforcement currently suffers from a lack of manpower and the geo-graphical conditions of isolation that obtain throughout much of the country (Zurick 2006).

6.4 Role of Forests in Supporting Livelihoods and Economy in the Region

The ecosystem services emanating from forests greatly contribute to human well-being in South Asia. While a large number of metrics have been developed in an attempt to estimate well-being, an approach used by the Millennium Ecosystem Assessment has been used here due to its close linkage with ecosystem services approach. The Millennium Ecosystem Assessments suggests 4 constituents of human well-being as shown in Fig. 6.4. These constituents include security, basic material for good life, health and good social relations which aggregately contribute to freedom of choice and action. The contribution of forest ecosystem services to each of these constituents of well-being is briefly discussed below in the South Asian context (MEA 2005).

6.4.1 Security

Mangroves forests are found in coastal areas of Bangladesh, India, Maldives, Pakistan and Sri Lanka pivotal role in protecting the human settlements from natural disasters like Cyclones and Tsunami. Cities like Kolkata in India are safeguarded by the mangrove covers at Sundarbans (see case studies on mangroves protected villages and reduced death toll during Indian super cyclone and Valuation of storm protection as an ecosystem function of mangrove in Sect. 6.5). Existence of forest helps allaying the extent of floods and helps considerably in its avoidance, a significant problem in countries such as India, Pakistan and Bangladesh. The high altitude forests existing in the Himalayan Range also help in the prevention of landslides in the area which can be catastrophic for the human settlements in the valleys. Thus, the role of forests in South Asia in reducing vulnerability to human existence is very significant. The role of forests is not only restricted to reducing the vulnerabilities but also in sustenance of life. The Himalayan range acts as a source of rivers which provides irrigation and drinking water to countries such as Pakistan and India and contributes to food, water and income security in the region (see case study of economic valuation of forests of Himachal Pradesh, India in Sect. 6.5). Since a major portion of agriculture in this region is dependent on rain for irrigation, forests act as sources of livelihood during crop failure and the non-agricultural season and thus act as a safety net. Timber and other building materials obtained from forests are used in building homes for the local community and tribes living inside and in vicinity to forest areas in the region.

A large population of South Asia is employed in agriculture, forestry and allied activities. While precise estimate of the magnitude of employment in forestry sector is difficult to obtain for South Asia, FAO states the following statistics on forests in South Asia.

6.4.2 Basic Material for Good Life

A large number of people in South Asia derive livelihood from forests, be it the collection of non-wood forest produces or extraction of timber. A large number of people are also involved in silviculture and associated forestry practices. Further there is significant number of people involved in various related industries from extraction to processing the wood as well as crafting to furniture making.

6.4.3 Health

The South Asian forests provide numerous services which are directly associated with the human health. This includes provisioning of fresh water for drinking, disease regulation or medicinal resources provided by forest.

6.4.4 Good Social Relations

Provisioning and regulating services provided by the forests have indirect effects on the social relations while they directly impact the material well-being of the people. These further improve the socioeconomic conditions of people where communities are formed, where people respect each other and develop social cohesion. Forest ecosystem services also include recreation services availed by the people.

6.4.5 Freedom of Choice and Action

This last component refers to the ability of individuals to take control of what happens to them and what they achieve. It is strongly mediated by other constituents of well-being as well as other factors e.g. education and socioeconomic conditions. For example, in many of the affluent countries with strong institutions, individuals can maintain freedom of choice and action even in the light of degradation forest ecosystem services. This however would not be possible for a poor dependent on forests for livelihoods living in a country with weak institutional structure (see case study of household characteristics and forest dependency in Nepal in Sect. 6.3). Similarly, reduction in provisioning of drinking water or fuel wood due to deforestation in many of the South Asian countries is known to have increase the time needed to collect some necessities, which in turn reduces time available for education, employment and others. The linkages between changes in ecosystem services and freedom of choice and action is however an understudied topic. This component is also a precondition for achieving other components of well-being, particularly with respect to equity and fairness. Significance of valuation of forest ecosystem services as a management tool.

Forests provide numerous ecological services which are critical to human existence. Many of the physical goods produced are tangible and thus have an economic value in the market since they are tradable. However, few goods and most of the non-tangible services provided by forests such as the oxygen we breathe, and which are perhaps even more important than the tangible goods provided by the forest, often do not have an economic value attached as they are not traded in market. This results in their undervaluation and poor management. Undervaluation has not only led to their overuse, but also misuse and abuse, as policy-makers and citizens do not get important signals that reflect their importance and scarcity.

Absence of any measure that reflect the economic value for these services from forests, various ecological services provided by forest are considered as free gifts of nature. The current National Accounting System of countries across the globe in general, and those in South Asia in particular reflects only the marketed value of few visible services supplied by forests. The economic value added by intangible services from forests as well as the consequences of degradation over time are completely ignored by the current system of national accounts. Lack of information on the value of forest ecosystem services is further reflected in terms of low budgetary allocations to the forestry sector in countries across South Asia.

With recent advances in forest database management and developments in the techniques of forest resource valuation it is now possible to demonstrate the immense contribution of forests to the growth and well-being of a country (see case study on the value of ecosystem services in Bhutan in Sect. 6.5). The Government of various countries of South Asia, in recognition of this have launched several processes to arrive at a better valuation of the various ecosystem services that our forests provide. This includes the calculation of Net Present Value (NPV) of using forest land for non-forestry purposes in India (see case study in Sect. 6.5), The Economics of Ecosystems and Biodiversity (TEEB), among various others.

A holistic and rigorous valuation of forests in this region in specific will highlight the importance of this natural resource which is otherwise treated as non-exhaustive resource and is severely exploited. Economic valuation also has the potential to increase awareness about the importance of forests in the well-being of human kind and ensure continued and sustainable productivity of this resource (see case study on willingness-to-pay for a national park in India in Sect. 6.5).

6.5 Case Studies on Valuation of Ecosystem Services

6.5.1 Mangroves Protected Villages and Reduced Death Toll During Indian Super Cyclone

Protection against coastal disasters has been identified as an important service of mangrove ecosystems. The impact of mangroves on human deaths during a 1999 super cyclone that struck Orissa, India was assessed and was found that villages with wider mangroves between them and the coast experienced significantly fewer deaths than ones with narrower or no mangroves (Das and Vincent 2009). Around 10,000 people were killed in the cyclone with 70% of them drowned. During this time due to increased anthropogenic activities the Mangrove area was reduced from 30,766 to 17,900 ha. Still the study shows how the mangrove cover of 17,900 ha was responsible in saving almost 265 lives (about 1.72 life in each village; 154 villages in total) during the super cyclone (Das and Vincent 2009). The statistical value of life came out to be US$190,000 (approx) so the total value for all the lives saves because of the presence of mangroves comes out to be US$50 million (Das and Vincent 2009).

6.5.2 Valuation of Storm Protection as an Ecosystem Function of Mangrove

The ecosystem services provided by mangroves are often ignored in the ongoing process of mangrove conversion. Services provided by the Bhitarkanika mangrove ecosystem in India and estimated cyclone damage avoided in three selected villages, taking the cyclone of 1999 as a reference point (Badola and Hussain 2005). The loss was assessed on the socio-economic status of the villages, the cyclone damage to houses, livestock, fisheries, trees and other assets owned by the people, and the level and duration of flooding. The loss incurred per household was greatest (US$153.74) in the village that was not sheltered by mangroves but had an embankment, followed by the village that was neither in the shadow of mangroves or the embankment (US$44.02) and the village that was protected by mangrove forests (US$33.31). The local people are aware of and appreciated the functions performed by the mangrove forests in protecting their lives and property from cyclones, and are willing to cooperate with the forest department in mangrove restoration (Badola and Hussain 2005).

6.5.3 The Value of Ecosystem Services in Bhutan

The total estimated value of the overall contribution of ecosystem services in Bhutan to human well-being is estimated to be US$15.5 billion/year (NU760 billion/year), significantly greater than the gross domestic product (GDP) of US$3.5 billion/year. 53% of the total benefits from Bhutan's ecosystem services accrue to people outside Bhutan (Kubiszewski et al. 2013). The remaining benefits accrue to people inside the country—15% at the national level, and 32% at the local level. Bhutan's combined per capita annual benefits are estimated at US $15,400/capita/year. Of this US$5000 is from goods and services captured in GDP and US$10,400 is from ecosystem services. These can be regarded as underestimates as it does not include other sources of benefits to people, including social and cultural values (Kubiszewski et al. 2013).

6.5.4 Willingness-to-Pay for a National Park in India

Contingent valuation method has been used to estimate the willingness to pay for the maintenance of a National Park in India i.e. Borivli National Park (BNP). Despite India being a developing country with medium to low income levels, it has been seen that people are willing to pay for preserving environmental amenities. Households are willing to pay exclusively for BNP, on average, Rs. 7.5, per month, for the next five years. Extrapolating to the city of Bombay, this amounts to a substantial US$325,575 (approx) each month for the next five years, or a present discounted value of in excess of US$16.28 million, which suggests a strong interest in environmental conservation (Hadker et al. 1997).

6.5.5 Economic Valuation of Forests of Himachal Pradesh, India

Total Economic Values for the forests of Himachal Pradesh, a State of India, are assessed in terms of services like direct consumptive benefits like timber, fodder, fuelwood, NTFPs; direct non-consumptive benefits like ecotourism & recreational and indirect benefits like watershed functions, carbon sinks, micro-climate, biodiversity, employment etc. (Verma 2000). The economic value of multiple contributions of forests to the economy should be given due weightage in allocation of funds. The actual forest cover in Himachal, covering an area of 14,346 km^2, generates economic value at the tune of US$12,100/ha and if the entire area under legal forests is used as denominator, the value reduces to Rs. US$4705/ha. The maximum per hectare value is generated by watershed function followed by carbon sink, biodiversity, ecotourism (all non-marketed values).The total economic value is 2.61 times the value of the growing stock, 980 times the total expenditure incurred in the forestry sector of Himachal Pradesh and 2607 times the revenue realized by the forests annually. This comparison proves gross underestimation of forestry sector's contribution in the economy of the state. When the GSDP of the state is corrected for Total Economic Value calculated through the current study the contribution of forestry sector increases from 5.26% of GSDP to 92.40% of GSDP. The total economic value of Himachal Pradesh's forests so estimated was approximately equal to US$21 million annually. Based on these estimates, in 2002, a special purpose vehicle called Compensation for Loss of Ecological Value (CLEV) was introduced in the state (Verma 2000).

6.5.6 *Net Present Value of Forest Diversion in India*

An agency seeking to use forest land for non-forestry purpose has to pay a charge called as Net Present Value to compensate for the loss of ecosystem services due to diversion of forests. In a recent study conducted to recalculate loss of ecosystem services due to forest diverted classifies forests into 14 Forest Type Groups and 4 Forest Canopy Cover Density Class. The array of forest goods and services valued in the study include timber, bamboo, fodder, fuelwood, NWFP, gene-pool conservation, carbon sequestration, carbon storage, soil conservation, water recharge, pollination and seed dispersal, and water purification. To estimate the duration for which the Net Present Value of losses need to be estimated, rotation period of dominant tree species was used. Further, it is found that about 50% of the total economic value of forests is accrued at the local level with 34% and 16% at the state and national level respectively. However, currently there is no mechanism to compensate stakeholders at local and regional level that incur losses due to forest diversion. To make NPV more site-specific, the study further suggests a premium on applicable NPV rates based on add-on factors of protected areas, hill talukas and forested wetlands (Verma et al. 2013).

6.6 Role of Forested Mountains in Provisioning of Water Downstream

The Himalayan Range in South Asia stretches from Afghanistan to Bhutan and has some of the highest peaks in the world such as Mount Everest and K2. The range is home to the world's highest peaks: the Eight-thousanders (peaks over 8000 m above sea level), including Mount Everest and runs from West-northeast to east-southeast in the shape of an arc extending over 2400 km. The wide variety of flora and fauna exists throughout the rage and differs according to the change in rainfall, altitude and soil conditions. The amount of yearly rainfall increases from west to east along the southern front of the range. It is also a home for diverse ecosystems and the services of these ecosystems is the source of life for not only for the mountainous population but also by the people living downstream. It has the third largest deposit of snow and ice after the Arctic and Antarctica. The Himalayan range encompasses about 15,000 glaciers, which store about 12,000 km^3 (3000 cubic miles) of fresh water. Its glaciers include the Gangotri and Yamunotri (Uttarakhand) and Khumbu glaciers (Mount Everest region), and Zemu (Sikkim) (IPCC 2007).

A number of perennial rivers originate from the Himalayas. Three of the world's major rivers, the Indus, the Ganges and the Tsangpo-Brahmaputra, rise in the Himalayas. While the Indus, and the Tsangpo-Brahmaputra rise near Mount Kailash, the Ganges rises in the Indian state of Uttarakhand. Their combined drainage basin is home to some 600 million people. This gives a fair idea on the water services provided by the Mountain ecosystem providing water not just for drinking but for irrigation and all other purposes. It won't be wrong to declare these rivers as the lifeline for the cities, towns and villages downstream, considering the criticality of drinking water supplied and increasing water scarcity in the region. Absence of these glaciers and the water services would likely lead to drought conditions in large part of India and Bangladesh. Moreover the rivers originating from the Hindu Kush Himlayas are also the main source of water for irrigation in the countries of India, Pakistan and Bangladesh. Looking at the dependency on these rivers, it can be deduced that these rivers are ensuring food and water security of the major chunk of South Asia (see case study of food and water security provided by Hindu Kush Himalayan region in Sect. 6.7). Hence the need to preserve this ecosystem becomes all the more important.

But due to the ever increasing anthropogenic pressure and the climate change, the mountain ecosystems in South Asia have been impacted in an unprecedented manner leading to increased vulnerability of these ecosystems. Though a number of organizations are working on the conservation of these ecosystems, huge efforts are required to deal with these problems. Some of these could include participatory natural resource management and transboundary cooperation for rejuvenation and long-term sustenance of these ecosystems.

6.7 Case Studies on Mountain Ecosystem

6.7.1 Sukhomajri—A Case of Forest and Watershed Protection in India

Sukhomajri is a village in the erosion prone Shivalik hills near Chandigarh in north Indian state of Haryana where there was a severe problem of erosion in the upstream catchment area leading to silting up of the downstream Sukhna Lake, a major recreation attraction for residents of Chandigarh (Kerr 2002). In 1977 four tanks were built to conserve rainwater. This increased water storage and provided protective irrigation to the rainfed crops, increasing yields. This became an incentive to the villagers to stop grazing and initiate afforestation and watershed protection in the catchment, to reduce erosion and prevent sedimentation of their tanks. As many of the grazers did not own land and thus would not benefit directly from the water and increased yields, an innovative step was taken—of splitting water ownership from land ownership. The water stored was made the property of all villagers—equally shared by both land holders and landless, thus the landed bought water beyond their share from the landless folks. The project gave increased yields in agriculture, protected forest cover, and reduced erosion in the Sukhna lake (Kerr 2002).

6.7.2 Food and Water Security Provided by Hindu Kush Himalayan Region

With limited land resources, inadequate energy supply, and growing water stress, South Asia faces the challenge of providing enough water and energy to grow enough food for the burgeoning population. Rice and wheat are the staple foods in South Asia; about 50% of dietary energy comes from these two crops. But these crops require huge amounts of water—about 1000 tonnes to produce 1 tonne of grain. Their production depends on the availability of water in the dry season and on irrigation facilities, which depend on water from the Hindu Kush Himalayas. In Pakistan, food, water, and energy security depends heavily on the state of the Indus River. The Indus irrigation system, the world's largest contiguous irrigation system, irrigates about 14.3 million hectares of farmland, representing about 76% of the cultivated area in Pakistan; it enables the production of more than 80% of the food grains of Pakistan and cash crops, in particular cotton. Agricultural water withdrawal in Pakistan is 170 billion cubic metres per year. Similarly, the Ganges River system is the main source of freshwater for half the population of India and Bangladesh and nearly the entire population of Nepal. The Ganges and Yamuna canal systems irrigate vast areas of India by using surface and groundwater received from the Himalayas. Almost 60% of India's irrigated area of 546,820 km^2 is in the Ganges basin. Water use for irrigation in the Ganges basin is about 100 billion cubic metres per year. The Brahmaputra River supports irrigation, hydropower, and fisheries for a vast part of Bangladesh, Bhutan, and India. Almost 6000 km^2 are irrigated using 1.4 billion cubic metres of water per year. The issues and challenges in the food, water, and energy sectors are interwoven in many complex ways and cannot be managed effectively without cross-sectoral integration (Rasul 2014).

6.8 Conclusion

Forest ecosystems in South Asia are important part of landscape and also support livelihoods of large number of rural populations. However, lack of economic value and their linkages with human wellbeing restricts decision-makers in recognizing their significance. In this chapter, we have highlighted the functions, value and management of forests by citing examples from various case studies spread in Indian sub-continent. We also address the important role of economic valuation in designing appropriate policy measures for balancing conservation and economic development in the region.

References

Adhikari B, Lovett JC (2004) Household characteristics and forest dependency: evidence from common property forest management in Nepal. Ecol Econ 48:245–257

Badola R, Hussain SA (2005) Valuing ecosystem functions: an empirical study on the storm protection function of Bhitarkanika mangrove ecosystem, India. Environ Conserv 1:85–92

Das S, Vincent JR (2009) Mangroves protected villages and reduced death toll during Indian super cyclone. Proc Natl Acad Sci 106:7357–7360

FAO (2012) South Asian Forests and Forestry to 2020 (Subregional Report of the Second Asis Pacific Forestry Sector Outlook Study). Food and Agriculture Organization of United Nations, Rome

Hadker N, Sharma S, David A, Muraleedharan TR (1997) Willingness-to-pay for Borivli National Park: evidence from a contingent valuation. Ecol Econ 21:105–122

IPCC (2007) Climate change 2007: Working Group I: The physical science basis. Intergovernmental Panel on Climate Change, Geneva

Kerr J (2002) Watershed development, environmental services, and poverty alleviation in India. World Dev 30:1387–1400

Khumbongmayum AD, Tripathi RS, Khan ML (2005) Sacred groves of Manipur, northeast India: biodiversity value, status and strategies for their conservation. Biodivers Conserv 14: 1541–1582

Kubiszewski I, Costanza R, Dorji L, Thoennes P, Tshering K (2013) An initial estimate of the value of ecosystem services in Bhutan. Ecosyst Serv 3:11–21

MEA (2005) Ecosystems and human well-being: a framework for assessment. Millennium ecosystem assessment. Island Press, Washington

Ninan N, Sathyapalan J (2003) The economics of biodiversity conservation-a study in a coffee growing region of India. In: International Association of Agricultural Economists Conference, August. Durban, South Africa

Poudel D, Johnson FH (2009) Valuation of crop genetic resources in Kaski, Nepal: farmers' willingness to pay for rice landraces conservation. J Environ Manage 90:483–491

Rasul G (2014) Food, water, and energy security in South Asia: a nexus perspective from the Hindu Kush Himalayan region. Environ Sci Policy 39:35–48

Rausser GC, Small AA (2000) Valuing research leads: bioprospecting and the conservation of genetic resources. J Polit Econ 108:173–206

Sarker S (2010) Mangrove: a livelihood option for the coastal community of Bangladesh. University of Chittagong—Institute of Marine Sciences and Fisheries, Chittagong, Bangladesh

Verma M, Negandhi D, Wahal AK, Kumar R (2013) Revision of Rates of NPV applicable for different class/category of forests. Indian Institute of Forest Management, Bhopal, India

Verma M (2000) Economic valuation of forests of Himachal Pradesh. In: Himachal Pradesh Forestry Sector Review. International Institute for Environment & Development, London

World Bank (2014) World Development Indicators. The World Bank, Washington, D.C.

Zurick D (2006) Gross national happiness and environmental status in Bhutan. Geogr Rev 96:657–681

Chapter 7
Urban Ecosystems: Functions, Value and Management

Jürgen H. Breuste, Muhammad Mushahid Anwar, Rab Nawaz and Mehwish Rani

7.1 Urban Ecosystem Characteristics—What Makes the Urban Ecosystem Special?

More than 50% of the mankind are living in cities and towns. This number is further growing. Urbanization is a multidimensional process that manifests itself through rapidly changing human population densities and changing land cover. The urban growth is due to a combination of two forces: natural growth and rural to urban migration. The last is the most important and responsible for the growth of many cities worldwide. The growth of cities forms large urban landscapes, particularly in developing countries. Since urbanization is a process operating at multiple scales, factors influencing environmental change in urban landscapes often originate far beyond city. Fluctuation in global trade, civil unrest in other countries, health pandemics, natural disasters, and possibly climate change and political decisions are e.g. factors driving social–ecological transformations of the urban landscape (Breuste et al. 2013:83) (Fig. 7.1).

Urban landscapes, often more extended above the administrative borders of a single city, are complex ecological systems dominated by humans. The urban ecosystems differ un several respects from natural ecosystems: in their climate, soil, hydrology, species composition, population dynamics and flow of energy and matter (Alberti 2009:1).

J.H. Breuste
Department of Geology and Geography, University of Salzburg,
Hellbrunnerstrasse 34, 5020 Salzburg, Austria

M.M. Anwar (✉) · R. Nawaz · M. Rani
Department of Geography, University of Gujrat,
Hafiz Hayat Campus, Jalalpur Jattan Road, Gujrat, Pakistan
e-mail: mushahid.anwar@gmail.com

H. Sandhu (ed.), *Ecosystem Functions and Management*,
DOI 10.1007/978-3-319-53967-6_7

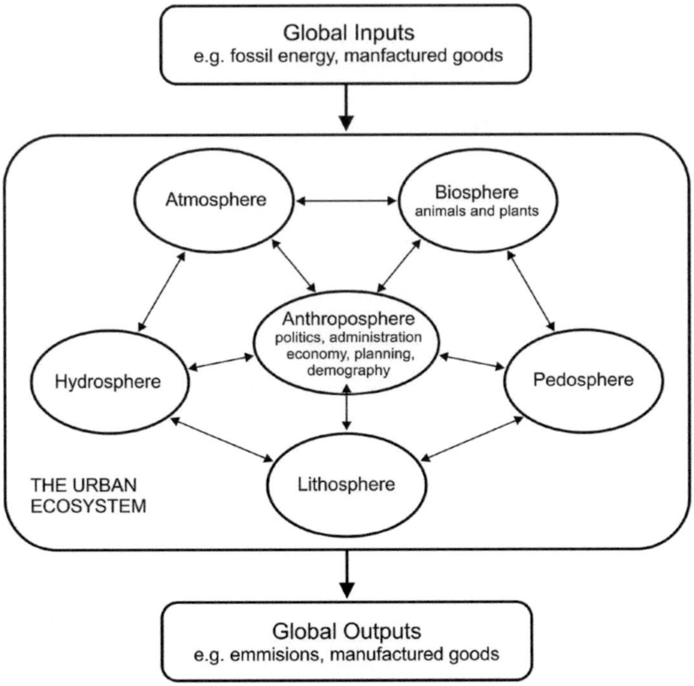

Fig. 7.1 Basic components of the urban ecosystem (Breuste, further developed from Marzluff et al. 2008:Viii; Breuste et al. 1998, design W. Gruber)

The foremost focus for research has often been on the negative impacts of the changes of urban ecosystems, particulary for human health and wellbeing, and how these can most effectively be mitigated by urban planning (Gaston 2010:35).

Urban Ecosystems

- are complex human phenomena,
- are social and biophysical phenomena,
- are a set of strongly interacting systems or spheres,
- are both drivers of, and driven by, ecological processes,
- have generalizable and definable internal structures, functions, and processes that produce cities' emergent properties. Many of those internalities are not yet identified or understood,
- have large impacts on Earth's ecological processes. Cities have huge ecological "footprints" caused by their needs for goods, energy and services and their capacity to import natural resources from, while exporting their emissions and wastes to, distant regions,
- also have the potential to offer unique opportunities for resource conservation and environmental impact mitigation (Marzluff et al. 2008:X).

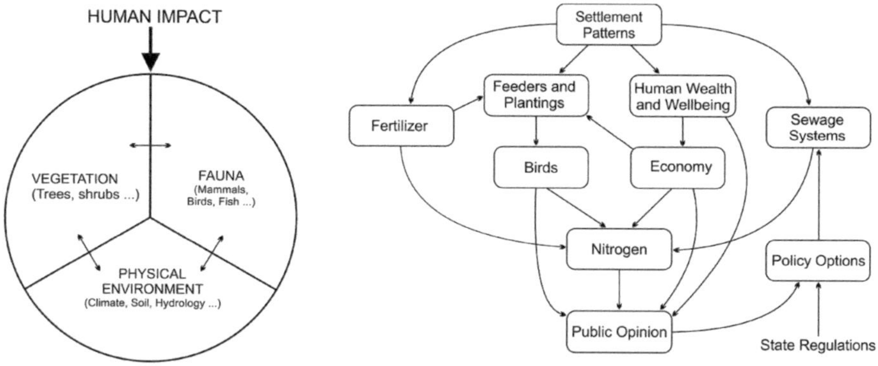

Fig. 7.2 Example of relationship in a typical study of ecology in the city (*left*) and ecology of the city (*right*) (Breuste, further developed from Marzluff et al. 2008:iX, design W. Gruber)

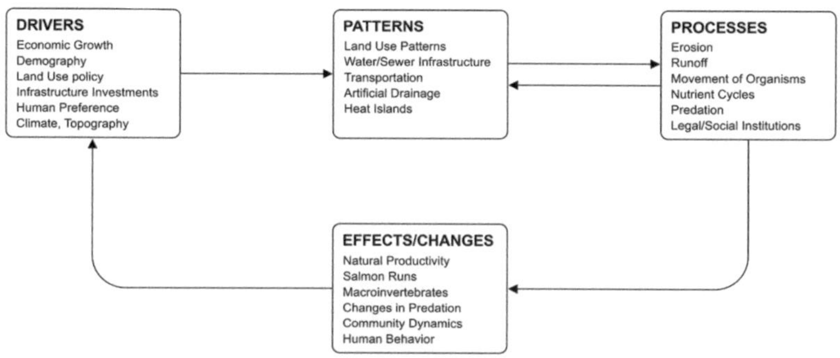

Fig. 7.3 Relationship between humans and nature. Abiotic (anthropogenic and otherwise) drivers cause patterns and processes that can be measured. They themselves affect interacting human and natural drivers by their effects and changes to the urban ecosystem (Breuste, further developed from Marzluff et al. 2008:X, design W. Gruber)

The relationship in urban ecosystems can be studied as ecology in the city and ecology of the city (McDonnell et al. 2009) (see Fig. 7.2).

Another option to study relationships in urban ecosystems is to investigate the relations between drivers, patterns, processes and effects/changes (Niemelä et al. 2011) (Fig. 7.3).

Marten (2001) shows the interaction between social system and ecosystem to a socio-ecosystem as urban ecosystem (Figs. 7.4 and 7.5)

Physical conditions that influence and determine fauna and flora in urban ecosystems as climate, soils and hydrology but also the overall steering 'physical' intervention of humans by the processes of land use and which create the new urban morphology. It is the human intervention by using the land for different purposes, changing the interventions/disturbance quality and quantity frequently and extends the influencing area which produce the conditions of urban ecosystems.

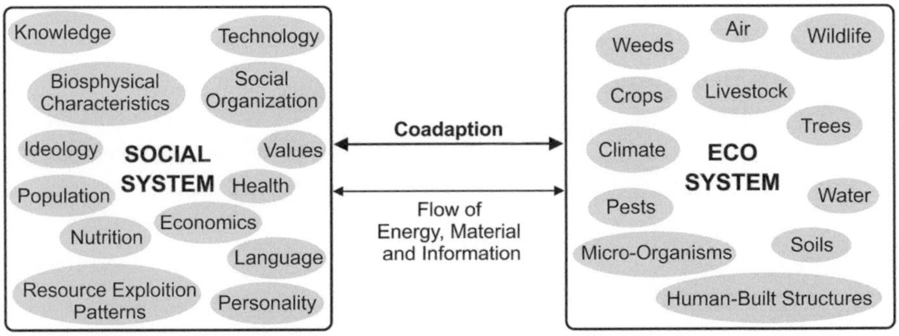

Fig. 7.4 Interaction of social system and ecosystem as urban ecosystem (Breuste, further developed from Marten 2001, design W. Gruber)

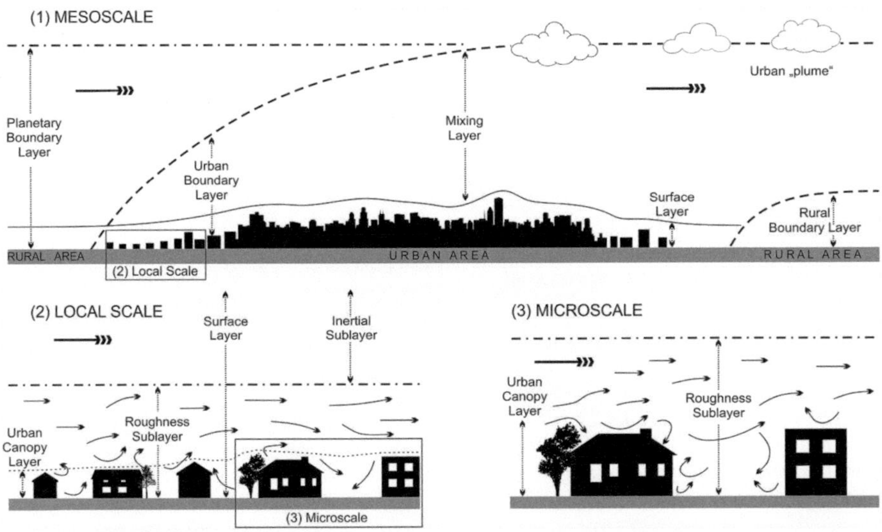

Fig. 7.5 Urban Ecosystem in different spatial scales, with an emphasis on atmospheric compontents (Breuste, further developed from Gaston et al. 2010:37, design W. Gruber)

With the replacement of the vegetation cover of land by paved surfaces and buildings the urban surface not only multiplies the former surface but introduces also instead of biologically working surfaces new surface materials like concrete, asphalt, stone plates, glass etc. These cover the soils and change their functions, have e.g. completely different thermal characteristics and influence all hydrological processes. Additionally the utilization activities introduce new matters, nutrients and pollutants into the natural system of water cycle and change the air quality. A new physical world is created—the urban ecosystem. But it is not only a new physical but also a new ecological one (Breuste 2009, 2011).

Land uses are particularly diverse, small scaled, intensive and influential not only on the targeted urban area. They consist of a mosaic of different, often very small-pieced land use types as basic elements and the key tool for undertaking applied urban ecological research and urban nature conservation (Breuste 2002). Urban structural or morphological units and types can be distinguished by their characteristic pattern of built and open spaces (Pauleit and Breuste 2011). It is the vegetation cover (as part of the surface cover) which is as component of mostly all urban land use types responsible for the ecosystem services which is everywhere needed and has to be extend targeted within urban development (Breuste 2011).

Most meteorological variables like temperature, heat stress, air pollution, wind etc. are influenced. New physical, biological, chemical characteristics, the urban heat island phenomenon and the impacts of urban climate on human health are the result. The urban climate system differs from non-urban/rural conditions in various aspects. These are high aerodynamic surface roughness (highly three-dimensional and therefore a very complex surface for all exchange processes with the urban boundary layer) which influences the vertical turbulence and wind field, different radiation and heat budget due to the physical properties of construction material like heat capacity and thermal conductivity and significant source of emissions from traffic and industrial sites as well as heating and air condition in terms of greenhouse gases, pollutants and direct heat release (Breuste 2011) (Figs. 7.6 and 7.7).

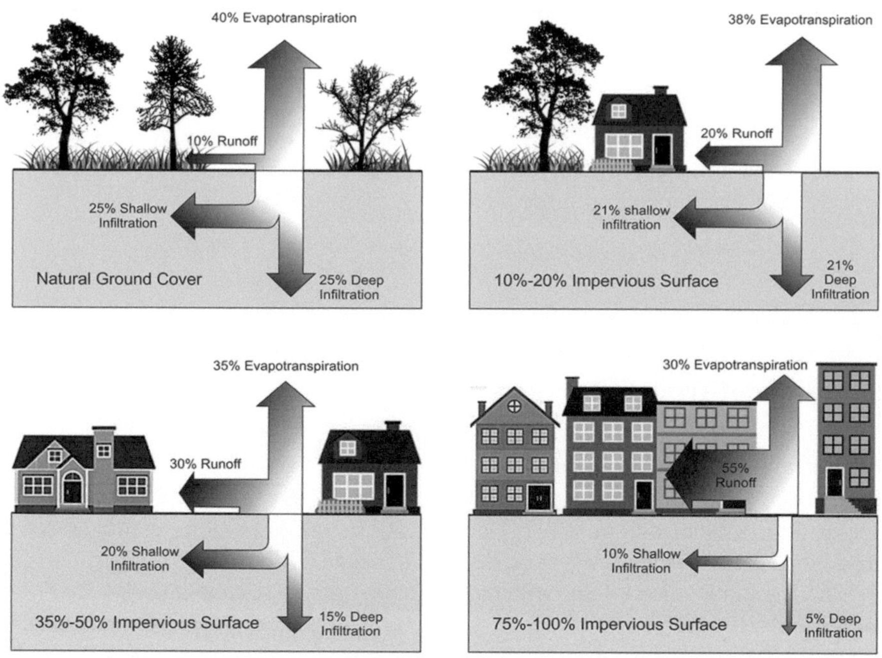

Fig. 7.6 Relationships between impervious surface cover and surface runoff, infiltration and evapotranspiration (Breuste, further developed from Gaston et al. 2010, design W. Gruber)

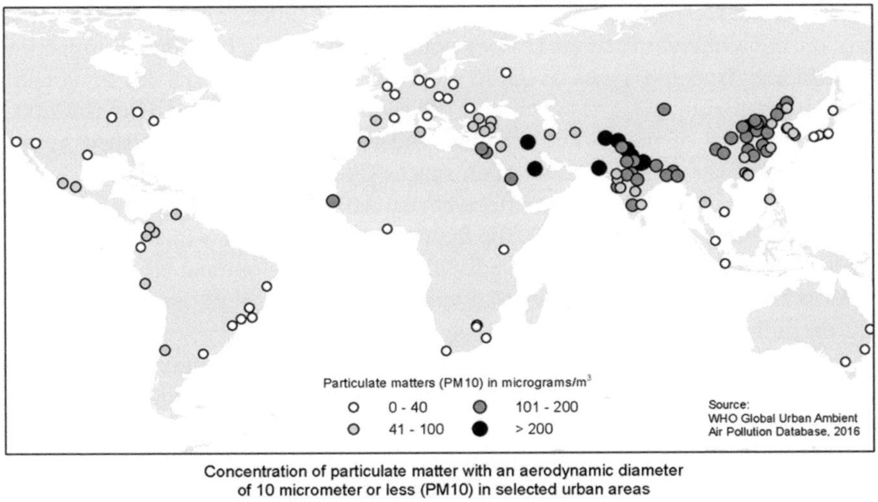

Fig. 7.7 Levels of particulate air pollution in 2016 within large cities over 1.5 Mio inhabitants. (Breuste, WHO 2016, design W. Gruber)

Land use characteristics explain the major components of the urban water cycle, together with the particular processes occurring between atmosphere, surface and subsurface. On the specific infiltration and runoff performance of urban surfaces latest innovations are presented. The stormwater retention capacities of paved and unpaved surfaces are specified from an urban water management perspective. Characteristic values of the particular water balance components on an annual basis as well as event-based runoff coefficients are outlined for several types of surfaces, reflecting the interactions between the atmospheric impact by rainfall and the hydrological phenomena on urban areas.

7.2 Urban Ecosystems Functions and Services—How Does the Urban Ecosystem System Work and How Can We Benefit from It?

Ecosystem Services (ES) describe how humans profit from ecosystem functions and processes without destroying them. The benefit can be economically or related to live quality (Costanza et al. 1997; DeGroot et al. 2002). The life quality stands in the centre of the Urban Ecosystem Service concept (UES).

With the publication of the Millennium Ecosystem Assessment (MEA 2005) the discussion on ecosystem services was more extended to urban ecosystems. Typically, urban ecosystem functions humans benefits from range from water and climate regulation functions over biodiversity and food to aesthetic and recreational services. According to McDonald (2009) UES are provided at different scales within an urban landscape: at the local (e.g. temperature regulation by tree shade, water and pollutant filtration at a single soil plot or food production in urban gardens and on agricultural land), at the regional or landscape scale (recreation, climate regulation, biodiversity) and at the global level (carbon mitigation, contribution to the continental or worldwide gene pool and biodiversity as such) (Breuste et al. 2013).

According to the Millennium Ecosystem Assessment (2005) and Costanza et al. (1997) we can define four categories of UES (Table 7.1):

- Provisioning services (food and timber production, water supply, the provision of genetic resources),
- Regulating services (regulation of climate extremes such as heavy rainfall and heat waves, floods and diseases, regulation of water flows, treatment and handling of waste),
- Cultural services (recreation and tourism, provision of aesthetic features, spiritual requirements) and, finally,
- Supporting services (soil formation and processes, pollination or energy, matter and nutrient fluxes).

Ecosystem properties are the set of ecological conditions, structures and processes that determine whether an ecosystem service can be supplied (van Oudenhoven et al. 2012). Human needs or demands convert a potential into an ecosystem service, since the services (and goods) are in fact currently valued, demanded or used. An ecosystem service is that's why only a service if there is a human benefit. The stakeholders, providers, users or beneficiaries of ecosystems and their services can be single persons, groups, or the society as a whole. They also in turn react upon ecosystems (Bastian et al. 2013) (Fig. 7.8).

Table 7.1 Services and indicators of quality of life related to the dimensions of sustainability

Sustainability dimension	Urban ecosystem service	Quality of life indicator
Ecology	Air filtration Climate regulation Noise reduction Rain water drainage Water supply Waste water treatment Food production	Health (clean air, protection against respiratory diseases, protection against heat and cold death) Safety Drinking water Food
Social sphere	Landscape Recreation Cultural values Sense of identity	Beauty of the environment Recreation and stress reduction Intellectual endowment Communication Place to live
Economy	Provision of land for economic and commercial activities and housing	Accessibility Income

Own listing according to Millennium Ecosystem Assessment (2005), Santos and Martins (2007), Breuste et al. (2013)

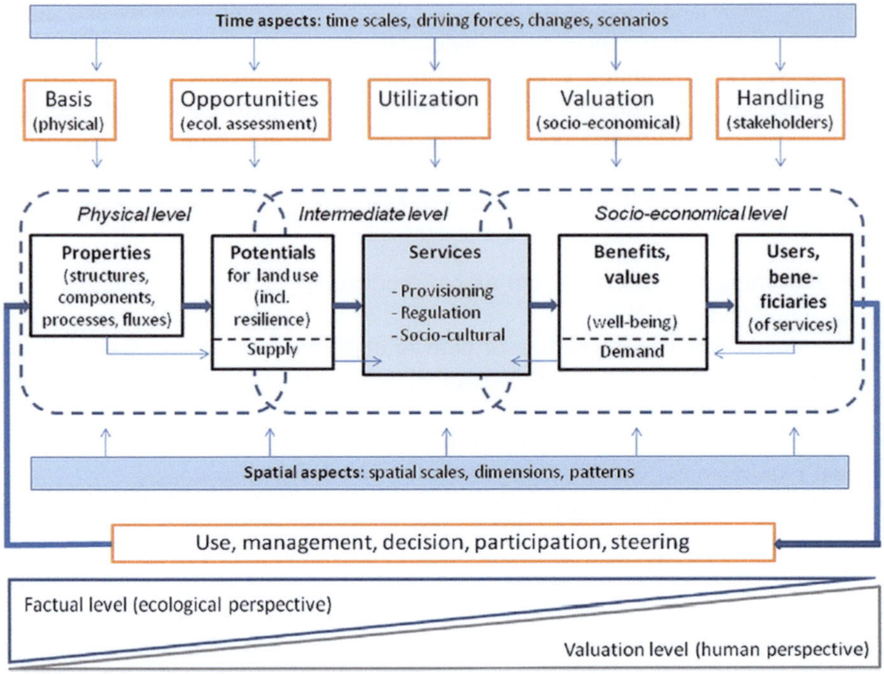

Fig. 7.8 Conceptual framework for the analysis of ecosystem services—the extended EPPS framework (Bastian et al. 2013)

Urban green areas are the providers of urban ecosystem services (Table 7.2). All kind of urban green areas can contribute as urban ecosystems (Gilbert 1989) to ecosystem services (Table 7.3).

7.3 Urban Ecosystem Management—How Can We Make Urban Ecosystems Better Working?

The main tasks of urban ecosystem management are reduction of harms and extension of benefits from ecosystem services. This needs to identify harms and benefits and to target them with management strategies and actions.

In many cities people suffer on:

- polluted drinking water
- polluted food
- polluted air
- noise
- thermal discomfort by high air temperatures in summer
- to less open space and green areas for recreation
- no nature contact

Table 7.2 Urban green as providers of ecosystem services (own listing according to Niemelä et al. 2010: 3229–3230)

Group	Ecosystem service	Service generating unit
Provisioning services	Timber products	Different tree species
	Food: game, berries, mushrooms	Different species in land, fresh-water and sea ecosystems
	Fresh water, soil	Groundwater infiltration, suspension and storage
Regulating service	Regulation of microclimate at the street and city level, changes in heating costs	Vegetation
	Gas cycles, O_2 production, CO_2 consumption	Vegetation, especially forests
	Carbon sequestration and storage	Vegetation, especially trees
	Habitat provision	Biodiversity
	Air pollution purification	Vegetation covered areas, soil microorganisms
	Noise cushioning in built-up areas and by transportation channels	Protective green areas, thick/wide forest, soft surfaces
	Rain water absorption, Balancing storm water peaks	Vegetation cover, sealed surface, soil
	Water infiltration	Wetlands (vegetation, microorganisms)
	Pollination, Maintaining floral populations, food production	Insects, birds, mammals
	Humus production and maintaining nutrient content	Litter, invertebrates, microorganisms
Cultural services	Recreation of urban dwellers	Biodiversity, especially in parks, forests and water ecosystems
	Psycho-physical and social health benefits	Forest nature
	Science education, research and teaching	Biodiversity

Table 7.3 Types of urban vegetation structures – influenced or created by urban land use (Breuste et al. 2013)

Vegetation Group	Vegetation structure type	Main urban ecosystem services and utilization	Main potential functions
(A) Vegetation remnants of the original natural landscape	Woods and forests	Recreation, BD	Timber production
	Wetlands	Biodiversity (BD)	Nature experience
(B) Vegetation of the cultural landscapes formed by agriculture	Meadows, pastures	Agriculture	Recreation, biodiversity
	Drifts, dry grasslands	Agriculture	
	Arable land	Agriculture	Recreation, biodiversity

(continued)

Table 7.3 (continued)

Vegetation Group	Vegetation structure type	Main urban ecosystem services and utilization	Main potential functions
(C) Ornamental, horticultural and designed urban vegetation spaces	Decorative green (flower beds, small lawn patches, bushes, hedges, etc.)	Decoration	Recreation, biodiversity
	Accompanied green along traffic lines or as addition to fill up the space between apartment blocks	Decoration	Recreation, biodiversity
	Gardens/parks	Recreation, decoration	Biodiversity
	Allotment gardens (territorially organized in allotment garden estates)	Recreation	Biodiversity
	Urban trees	Decoration	Biodiversity
(D) Spontaneous urban vegetation (areas)	Spontaneous herbaceous vegetation	None	Biodiversity, nature experience, recreation
	Spontaneous bush vegetation	None	Biodiversity, nature experience, recreation
	Spontaneous pre-forest vegetation	None	Biodiversity, nature experience, recreation

Die pollution of environment (water, air, soils, food etc.) must be solved on source by cleaning technologies. Urban Ecosystems can only partly support this (e.g. cleaning capacity for air pollution or water by vegetation, noise reduction by tree plantations etc.).

It remains the demand for thermal regulation, recreation and nature contact in the residential areas and in the whole cities as a demand and need which can't be technically managed. The urban ecosystem management must target to identify, managed and secure those urban ecosystems providing:

- food (fruit and vegetable production)
- recreation, including contact to different forms of nature (from forest to garden)
- climate regulation

These services are only provided by "green" and "blue" areas, vegetation and water bodies. They have to be secured, especially when they are in public hands and created to a network of green (and blue) infrastructure. The urban management must target to the services, what is actually only rarely the case, especially not in many fast growing cities in developing countries.

The management of services (often regulated and controlled by decisions and legislation tools) can modify or change the properties and potentials of ecosystems. Appropriate urban ecosystem management has to bridge the gap between the present state and future targets for urban ecosystem services. Urban ecosystem management means treats areas at various scales in such a way that urban ecological services and biological resources are restored and conserved, while appropriate human uses are sustained (Schneiders et al. 2012; Bastian et al. 2013).

Mismatches between spatial and temporal scales of ecological process on the one hand, and social scales of monitoring and decision making on the other have not only limited our understanding of ecological processes in urban landscapes, they have also limited the integration of urban ecological knowledge into urban planning. Human processes and cultures are fundamental for sustainable management of urban ecosystems, and in urban planning it is becoming more and more evident that urban management needs to operate at an ecosystem scale rather than within the traditional boundaries of the city. Urban ecosystems provide important large-scale probing experiments of the effects of global change on ecosystems, since, for example, significant warming and increased nitrogen deposition already are prevalent and because they provide extreme, visible, and measurable examples of human domination of ecosystem processes.

Urban ecosystem services are still mostly no clear target of urban management. The approach of "green and blue services" (Westerink et al. 2002) or "sustainable land use governance" (Ravetz 2000) are new integrative concepts of urban planning and urban governance. Necessary integrative planning means the removal of sectorial thinking and actions in favour of more holistic concepts of regional and local regeneration and adaptation (Ravetz 2000). Accordingly, expert-driven formal planning must be enhanced and accompanied by participatory processes (Elmqvist et al. 2008; Breuste et al. 2013:103).

7.4 The Sub Continent and Urbanizing World

Asia is urbanizing and there is growing proportion of urban population of its total population. South Asian economies are passing through transition phase today. These countries are rapidly transforming into manufacturing and services economies, leading to urban population growth with every passing day. Countries in Indian-subcontinent were primarily rural, particularly India, Bangladesh and Pakistan at the time of independence. Generally it is still a rural region and is passing through a phase of rapid transition. More and more people are moving daily from rural to the urban areas in the quest of basic facilities and better life opportunities including education, health, employment, recreation, transport and many more. Urban population of South Asia has risen to 30.1% of the total population in 2010, from 15.6% in 1950 (Mathur 2011). According to United Nations (UN 2010) there are three potential factors, which contribute to the urbanization trend in Asia. These factors are given as under:

i. Net inward migration to urban areas:

Number of people moving from rural to urban areas is higher than number of people moving from urban to rural areas. As a result there is net inward migration to urban areas leading to urbanizing Asia. It is a significant factor in most Asian countries.

ii. Natural urban population growth:

Natural population growth is higher in urban areas than in rural areas. This factor is rarer, but also contributes to urbanization trend in countries.

iii. Reclassification of settlements:

There is reclassification of rural settlements as urban settlements, increasing proportion of urban population. As cities grow outwards, urban areas envelop relatively populous rural areas. As the process of urbanization in South Asian region is ungoverned, its cost is much higher and is affecting economies in two

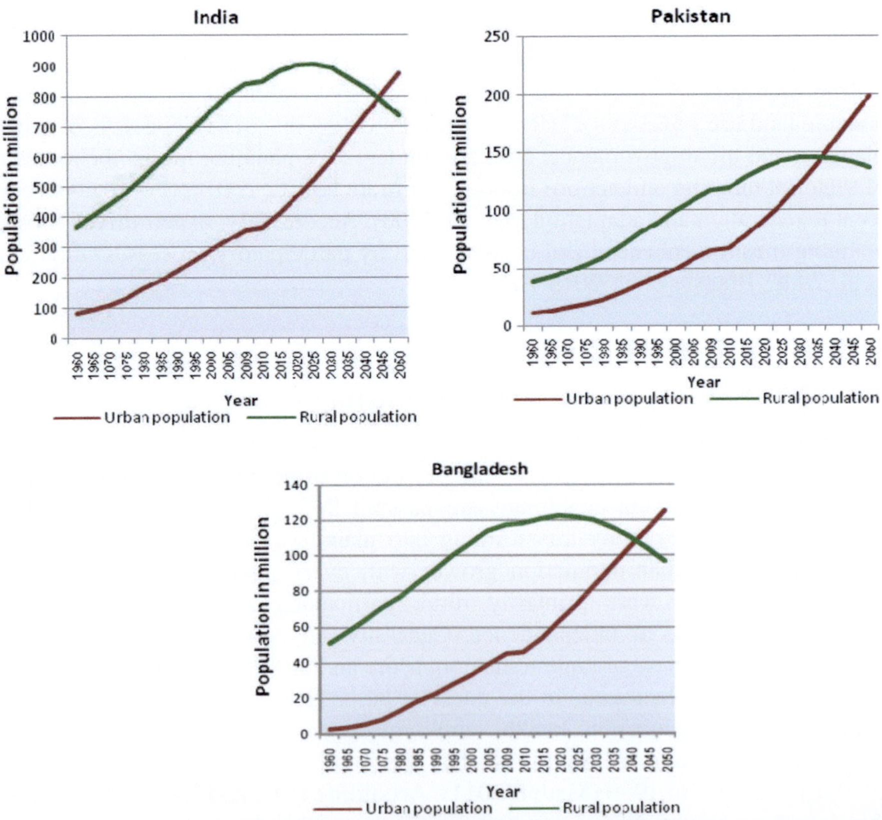

Fig. 7.9 Urban transition in sub-continent economies. *Source* Mathur (2011)

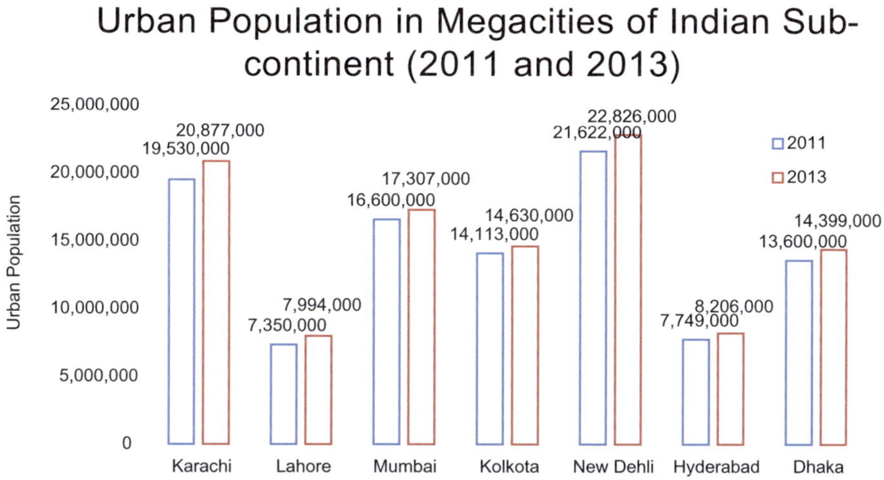

Fig. 7.10 Urban population in megacities of Indian-continent. *Source* Demographic World Urban Areas (2013, designed M.M. Anwar)

Table 7.4 Urban population growth

Country	Years				
	1985–90	1990–95	1995–2000	2000–2005	2005–10
India	3.15	2.88	2.63	2.35	2.39
Pakistan	4.29	3.27	3.26	3.82	3.04
Developing countries	3.75	3.19	2.96	2.68	2.53

Source World urbanization prospects. The 2007 Review regrading Asian countries from 1985 to 2010 (design M.M. Anwar)

ways. Firstly, 'rural-push' factor give rise to rural unemployment and increased poverty. Secondly, urbanization generates exceptional financial and management challenges in the urban system (Dutt and Noble 2004) (Figs. 7.9 and 7.10; Tables 7.4 and 7.5).

The rapid development and growth of cities is also common demographic phenomenon in Pakistan, which resulted to an increased degree of urbanization. The level of urbanization has increased from 17.4% in 1951 to about 32.5% in 1998 and it reached 36.2% in 2011 (Arif and Hamid 2009; CIA 2013). Karachi is 7th largest city of the world, with a population of 20,877,000. This rapid urban growth of megacities causes severe social, economic, ecological and management issues, which create barriers for sustainable development.

According to the World Bank, South Asia accounts for 46% of the world's urban poor on a poverty line of US$ 1.08/day, and 39% if the poverty line is changed to US$ 2.15/day.2 The relatively high GDP growth registered in recent years by several South Asian countries does not seem to have been translated into any

Table 7.5 Average annual rate of population change in south Asian countries

Country	Percentage rate of change			
	2000–2005		2005–2010	
	Total	Urban	Total	Urban
Bangladesh	1.89	3.58	1.67	3.45
Bhutan	2.63	6.60	1.43	4.88
India	1.62	2.35	1.46	2.39
Maldives	1.57	5.62	1.76	5.29
Nepal	2.08	5.27	1.97	4.87
Pakistan	1.82	2.82	1.84	3.04
Sri Lanka	0.43	-0.31	0.47	0.46

Source World urbanization prospects. The 2007 Review regrading average annual rate of population change in South Asian countries from 2000 to 2010 (design M.M. Anwar)

noticeable improvement in the quality and quantity of urban infrastructural services Mathur (2011).

South Asia's urban population—currently placed at 486 million persons—has risen steadily at an annual exponential rate of 2.9%. The level of urbanization as indicated by the proportion of total population living in the urban areas has risen from 15.6% in 1950 to 30.1% in 2010. Annual increments to urban population have been of the order of 9–10 million and are expected to increase to 14–15 million in the next decade. Although South Asia is among the least urbanized of the seven regions, its urban significance in global urban hierarchy, measured in terms of its share in world's total urban population, has progressively risen. Rural-urban migration has contributed significantly to the process of urbanization in subcontinent countries. The pattern of urbanization in South Asia is marked by a growing number of mega-cities. According to the United Nations, Dhaka's population which currently stands at 13.5 million will increase by 31% over the 2010–2020 period, and that of Karachi by 29%. Mumbai is projected to attain a population of 24 million by 2020 Mathur (2011).

At the moment, **India** is among the countries of low level of urbanization. Number of urban agglomeration/town has grown from 1827 in 1901 to 5161 in 2001. Number of population residing in urban areas has increased from 2.58 crores in 1901 to 28.53 crores in 2001. Only 28% of population was living in urban areas as per 2001 census (Datta 2006).

Pakistan's current population is about one-third urban. Much of Pakistan's urbanization is driven by migration. In past decades, Indian Muslims and Afghans fled to Pakistani cities to escape war back home. Today, rural Pakistanis are entering cities to escape war, insecurity and natural disasters, and also to seek new livelihoods and better basic services. The large natural increase in Pakistan's total population also explains the country's rising urban population. (Kugelman 2013) (Table 7.6).

Poverty is urbanizing rapidly in the developing world; indeed, some observers believe that poverty is now mainly an urban problem. This urbanization of poverty—by which we mean a rising share of the poor living in urban areas—has been viewed in very different ways by different observers. To some it has been seen as a positive

Table 7.6 Urban Population a projection

Country	Urban population			AAGR[a]		Percentage of urban to total population		
	In million			In percentage				
	2000	2010	2020	2000–10	2010–20	2000	2010	2020
Bangladesh	33.21	46.15	62.89	3.29	3.09	23.59	28.07	33.89
Bhutan	0.14	0.25	0.35	5.80	3.36	25.00	35.21	42.68
India	288.43	364.46	463.33	2.34	2.40	27.66	30.01	33.89
Maldives	0.07	0.12	0.19	5.39	4.60	25.93	41.38	52.78
Nepal	3.28	5.56	8.74	5.28	4.52	13.43	18.63	24.78
Pakistan	49.09	66.32	90.2	3.01	3.08	33.14	35.90	39.88
Sri Lanka	2.97	2.92	3.36	−0.17	1.40	15.82	14.31	15.48
Total urban population	377.19	485.78	629.06	2.53	2.58	27.42	30.08	34.24
Developing countries	1968.2	2556.47	3188.1	2.62	2.21	40.00	45.08	49.76
Asia	1360.9	1757.31	2168.8	2.56	2.10	36.80	42.17	47.19

[a]Average annual growth rate
Source Mathur (2011), urban population projection from 2000 to 2020 (design M.M. Anwar)

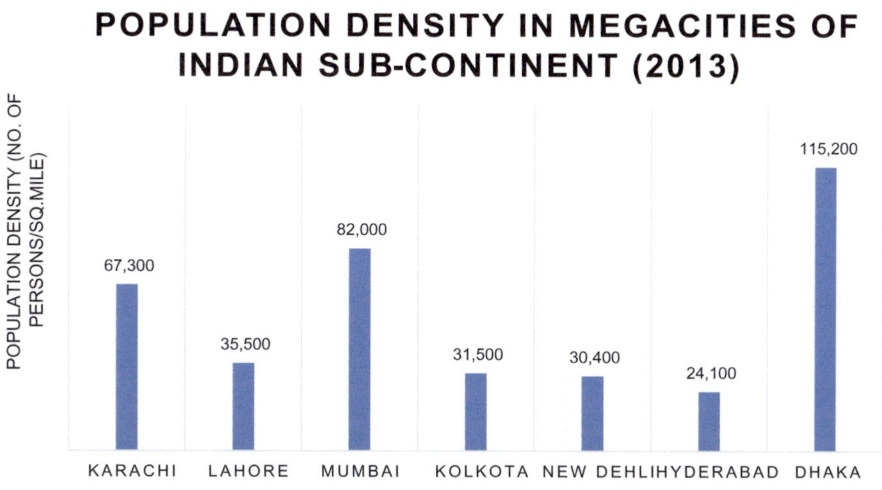

Fig. 7.11 Population density in megacities of Indian-continent. *Source* Demographic World Urban Areas 2013 deigned M.M. Anwar

force in economic development, as economic activity shifts out of agriculture to more remunerative activities, while to others it has been viewed in a less positive light—a largely unwelcome forbearer of new poverty problems" At the same time, the numbers of the urban poor have risen from 113 million in 1993 to 134.8 million in 2002, an increase of 19% over a ten-year period Mathur (2011) (Fig. 7.11).

7.5 The Most Dynamic Megacities on the Indian Sub-continent—Karachi, Mumbai, Calcutta, Dhaka, Lahore, New Delhi, Hyderabad

Urbanization is a dramatic form of highly irreversible land transformation. This is a major change worldwide but it is exceptionally dynamic in Indian Sub-continent. In 2007, world's urban population reached 3.3 billion and is expected to rise. It has been estimated that 60% of the world's total population will be urbanized by 2030. This rapid urbanization is caused by attraction of opportunities of employment and economic development (FIG 2010). The largest category of urban agglomerations is known as 'Megacity'. According to United Nations (UN 2007), megacities are municipalities having population more than 10 million. Today, there are 28 megacities worldwide and India has three dynamic megacities including Mumbai, Delhi, and Kolkata. In 2020, three other cities (Bangalore, Hyderabad and Chennai) will become megacities and the country will have largest number of megacities in the world (Rück 2005) (Fig. 7.12).

7.5.1 Pakistan

Environment sustainability in Pakistan is threatened due to unsustainable pattern of economic development. Problem connecting to environment are including high population growth rate, lack of public awareness and education, mismanagement of water and other natural resources as well as unplanned urban and industrial expansion. Economic growth is also hampered by environmental degradation and depletion of natural resource base. The city of Karachi alone contributes 21.7% to

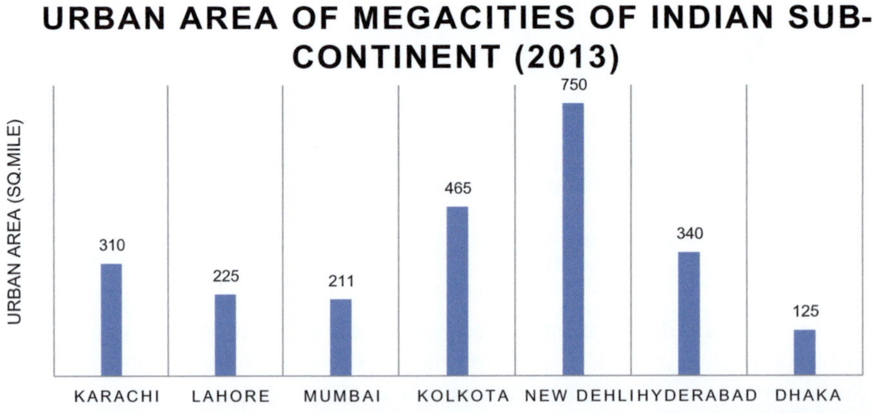

Fig. 7.12 Urban area of megacities of Indian-continent. *Source* Demographic World Urban Areas (2013, designed M.M. Anwar)

Table 7.7 Population distribution in Lahore municipal authority (LMA) (1998–2001)

District	Population			
	1998	2001 growth	Sf	Percentage of total population
Lahore	6,318,745	7,041,839	3.399	91.28
Seikhupura (share of LMA*)	499,012	541,682	2.56	07.02
Kasur (share of LMA)	120,838	131,097	2.54	01.70
LMA	6,938,591	7,714,618	3.32	100.00

Source Master Plan for Lahore 2004 regarding population Distribution in Lahore, designed M.M. Anwar

Table 7.8 Urbanization (1951–1998) in Lahore district

Census year	Population		Proportion of urban population (%)
	Lahore district urban	Lahore district	
1951	861,279	1,134,757	75.90
1961	1,312,495	1,625,810	80.73
1972	2,189,530	2,587,621	84.62
1981	2,988,486	3,544,942	84.30
1998	5,209,088	6,18,745	82.44

Source Master Plan for Lahore 2004 regarding urbanization in Lahore, designed M.M. Anwar

the total urban population of Pakistan, while the city of Lahore contributes 12.7%. They along with the biggest seven cities including Faisalabad, Rawalpindi, Multan, Hyderabad, Gujranwala, Peshawar and Quetta contain 54.6% of Pakistan urban population (Sattar 2012).

Pakistan's population in mid 2004 was estimated as 149 million—which is 1.9% higher than the previous year. According to the Population Census 1998, the population of **Lahore Metropolitan Area** (LMA) was 6.94 million which is estimated to have grown to 7.71 million in 2001. More than 91% of this population (7.04 million) lives in Lahore City District, 7.0% (0.54 million) in Tehsil Ferozewala of District Seikhupura and 1.7% in parts of Tehsil Kasur of District Kasur (Bajawa et al. 2005) (Tables 7.7 and 7.8).

7.5.2 *Urbanizing Problem in Lahore*

Urbanization phenomenon occurred because of **migration** (**Rural to Urban and International**) and natural increase. Natural increase is another factor which has also increased the demand for housing in the country. Both these factors (natural increase and migration) exerted a huge pressure and resultantly the population of Lahore increases very rapidly. Issues addressed include migration to urban centers, the immediate environmental and health impacts of urban pollution on developing country cities, and the link between crime and security. Urbanization does not have only local environmental impacts but also large so-called '**ecological footprints**' beyond their immediate vicinity. Intensive and extensive exploitation of natural resources to support urban economy includes excessive extraction of energy resources (including fuel wood), quarrying and excavation of sand, gravel and building materials at large scales, and over extraction of water. Other effects can be felt further afield such as pollution of waterways, long-range air pollution that impact on human health as well as on vegetation and soils at a considerable distance (Henderson 2006).

Taha (2010) presented trends in the population growth and urbanization in Lahore with the growth in the air pollution, water pollution and change in the forestation recourses in Lahore district. Air pollution has always been a problem in Lahore. About 2% of total deaths are attributed to air pollution in Lahore. According to this report 22,700 people die because of increase in urban air pollution. This inferior air quality in Lahore is the effect of vehicular and industrial emissions. Clean drinking water is scarce in Lahore. Most of the population in Lahore never drinks clean water.

Species threatened by urbanization also tend to be threatened by agriculture, recreation, roads and many other human impacts, emphasizing the uniquely far reaching transformations that accompany urban sprawl (Joshua and Ali 2011) (Tables 7.9 and 7.10).

Allama Iqbal Town: The feeding station was left in the Asif Block of Allama Iqbal Town area for a week. Allama Iqbal Town has an area of 387,437 m^2 and a population of 567,000 (Population Census Report 1998). The absence of graminivorous birds in the area except for a few House Sparrows was possibly because of the kitchen scraps given to the birds which attracted crows and mynas which possibly scared away the major avian diversity. The lack of roosting sites in the form of vegetation, however, was remarkable.

Data Ganj Bakhsh Town (Mall road opposite high court): With a total Population of 712,000 persons according to the 1998 census and an area of 28,110 m^2. Data Ganj Baksh ranks the second highest in congestion i.e. 25.33 persons per m^2. It is one of the smallest towns of Lahore, equal in size to Gulberg Town and larger in size only to Samanabad Town. No birds were observed at the feeding station till 6:30 a.m., this was strange, as previous studies on estimating bird Density state that the optimal time to count birds was during the first 4 h after sunrise. Many House Swifts were observed flying with their mouth open over small shrubs. This behaviour was also observed in Aziz Bhatti Town and Gulberg Town.

Gulberg Town (Gulberg III): Area wise Gulberg Town and Data Ganj Baksh Town were similar in size, (28,110 m^2) however; a total population of 571,000 persons according to the 1998 determines congestion level, 20.31, which is less than that at Data Ganj Baksh Town. Contrary to observations at Aziz Bhatti Town, birds in this area fed on all types of seeds. Many laughing doves and Ring necked Doves were observed sitting under the table but none on the table. The number of birds visiting the site decreased in the afternoon. Birds had consumed all types of

Table 7.9 Number of birds visiting the feeding station in the summer season (July 2010–September 2011)

Name of town	House sparrows	Laughing dove	Blue rock pigeon	Silver bill	Common babbler	Common myna	Ring necked dove	Rose ringed parakeet	Red vented bulbul	Bank myna
Aziz Bhatti Town	169	7	5	0	0	0	1	0	0	0
Nishtar Town	375	5	0	29	0	0	0	0	0	0
Gulberg Town	1266	19	0	0	35	0	0	0	0	0
Wagah Town	217	0	0	0	0	0	0	0	0	0
Data Gunj Baksh Twon	59	18	0	8	0	0	0	0	0	0
Allama Iqbal Town	5	0	0	0	0	5	0	0	0	0
Shalimar Town	110	0	0	0	0	0	0	0	0	0
Ravi Town	239	0	8	0	0	0	0	0	0	0
Samnabad Twon	289	13	17	0	0	0	0	0	0	0

Source Joshua and Ali (2011, designed M.M. Anwar)

Table 7.10 Number of birds visiting the feeding station in the winter season (December 2010–February 2011)

Name of town	House sparrows	Laughing dove	Blue rock pigeon	silver bill	Common babbler	Common myna	Ring necked dove	Rose ringed parakeet	Red vented bulbul	Bank myna
Aziz Bhatti Town	122	2	0	0	0	7	0	0	0	0
Nishtar Town	2164	5	0	0	0	2	4	11	32	0
Gulberg Town	1164	30	0	0	16	0	2	0	0	0
Wagah Town	664	1	0	0	0	4	0	0	0	11
Data Gunj Baksh Twon	258	4	0	18	0	5	0	0	7	0
Allama Iqbal Town	12	0	8	0	0	0	0	0	1	0
Shalimar Town	241	1	0	0	0	0	0	0	0	0
Ravi Town	985	3	35	0	0	0	0	0	0	0
Samnabad Twon	310	11	2	0	0	0	0	0	0	0

Source Joshua and Ali (2011, designed M.M. Anwar)

Fig. 7.13 Seeds set out in different quadrants of the Feeding station. *Source* Joshua and Ali (2011)

seeds however the seed of least preference was again Chick Peas, Mustard Seeds and Corn. Source: Joshua and Ali (2011) (Fig. 7.13).

Wagah Town (**Batapur**): Wagah Town was the third largest town of Lahore with the population congestion of 1.325. This explains the largest diversity of birds present in the area. The number of birds that visited the feeding station was however very low especially in the summer. In the winter season, however, the case was different as the crops were not in the seeding season and the number of insects available during the summer season had decreased as the season changed. The number of House Sparrows that foraged at the feeding station in the winter season increased three fold. Source: Joshua and Ali (2011).

7.5.3 Karachi

It associated with land attributes and human activities expresses a city's social and economic functions. Green spaces in the form of species composition, which performance echoes ecological and environmental functions. The green spaces can be denoted by urban factors, such as public parks, neighborhood parks (Anwar 2006).

Karachi is now among the ten top ranking largest cities in the world. Even at decreasing average annual growth rate (from 4.15% in 2005 to 3.5% in 2020), the increase in absolute terms is staggering and will put heavy pressure on the physical, infrastructure, financial and institutional systems of the city.

A large segment of **Karachi's population**, roughly 40%, is afflicted with poverty. As these impact the environment, ecosystem and growth potential of the city. Karachi's population is diversified in terms of ethnicity and economic conditions. Apart from in-migrants from Pakistan's provinces, a large number of migrants from Afghanistan, Bangladesh and other South Asian countries have settled in the city. CDGK (City District Government Karachi 2010–11) (Fig. 7.14).

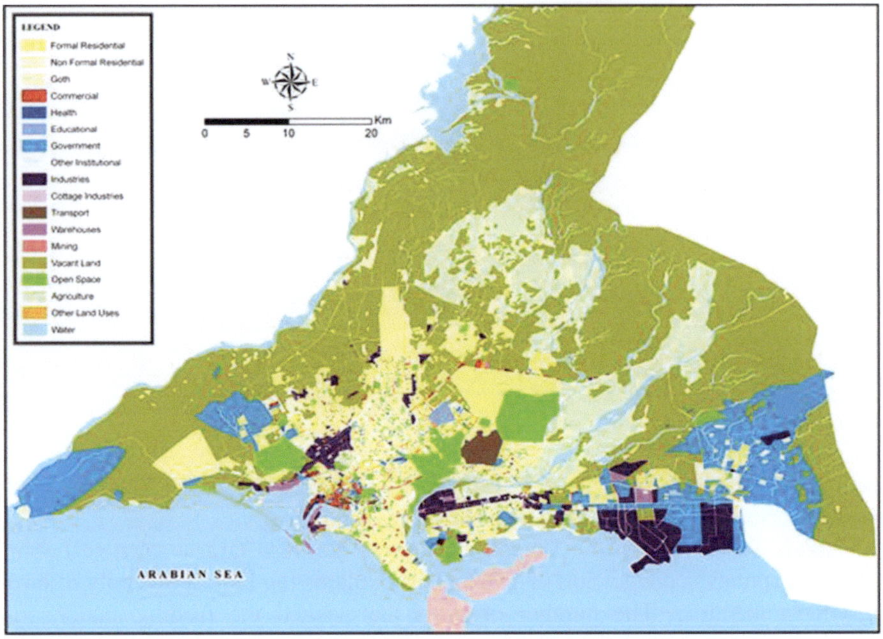

Fig. 7.14 Karachi existing land use—2005. *Source* (CDGK "Current Conditions and Issues" 2010–11)

With expansion of the built-up area and continuous densification the built environment has been marred by intense congestion, lack of cleanliness, unsanitary conditions, and poor maintenance of public infrastructure, over construction and enormous encroachment of foot-paths, streets, roads and public amenity open spaces. Water quality and water pollution are important environmental concerns. Most water available in Karachi does not meet the water quality standards of WHO (CDGK "Current Conditions and Issues" 2010–11).

7.5.4 Urbanizing Problems in Karachi

Air pollution is a serious environmental problem in the Karachi city. Automobile exhaust, industrial emissions, open burning of garbage, domestic and commercial fuel sources cause high increase in air. **Solid waste** from industries (also dumped outside the factory premises) is burnt in incinerators not always designed for hazardous waste disposal, like chemicals, pesticides and hospital waste. The Lyari and Malir rivers, which have been converted into large **open sewers**, cause severe marine pollution along the coast and impact the harbor's **marine environment**. Further, solid waste dumping and hazardous industrial sewage aggravates the contamination of the Lyari River. Institutional framework for environmental control is weak and indecisive with overlapping responsibilities (CDGK "Current Conditions and Issues" 2010–11). Ships discharge pollutants directly into the marine environment of the harbor. Human consumption of such sea-food may cause anemia, kidney failure and brain damage.

The **polluted marine environment** has also damaged the ecological balance maintained by the mangrove ecosystem that provides a natural habitat for inter-dependent communities of invertebrate's fish, shrimps, crabs, birds and reptiles. Protection of the mangrove ecosystem and minimizing pollutants discharge into the sea requires immediate corrective measures to preclude the fast degradation of the marine environment (CDGK "Current Conditions and Issues" 2010–11).

The **Karachi's coastline**, with its fine beaches, backwaters, mud flats and mangroves ecosystem, is a natural asset. Its scenic attraction and potential for recreation activities provide the basis for select urban development that combines the natural advantages, social and economic growth while promoting sustainability. There is considerable scope for development of several activities including commercial-amusement complexes and waterfront architecture. **Coastal development** is a special area requiring a plan carefully integrated with the overall metropolitan planning and has to be based on well-defined land use and urban design principles (CDGK "Current Conditions and Issues" 2010–11).

7.5.5 India

India's urban population is the second largest in the world, after China. Both, natural increase and rural to urban migration have contributed towards the urban growth in India. Component of urban growth has attributed to mainly three components and the largest contributor is natural growth 41.7% during 1971–81 and 58.7% during 1981–91. Almost the similar trend has been found by (Pathak and Mehta 1995) Agriculture, industry, energy production, urban development, grazing, logging, mining and other land uses relate to human activity or economic function associated with a specific piece of land. Land use is a product of interactions between cultural backgrounds, state and physical needs of the society with the natural potential of land (Karwariya and Goyal 2011).

Often improper urban land use is responsible for various forms of environmental degradation and ecological disturbance. It is essential to know the natural characteristics, extent and location, its quality, productivity, suitability and limitations of urban uses for sustainable utilization of the land ecosystems. The need for new housing, schools, industries, transportation and other civic amenities increases with increase in population. The sprawling process of expansion is disordered, unplanned, leading often to inefficient and unsustainable urban expansion patterns (Travisi and Camagni 2005).

In Delhi, India built-up area is 16.86% of the total city area of 1490 km^2 during 1997–2008. This area mainly came from agriculture land, waste land, scrub-land, sandy areas and water bodies. In comparison the increment in forest cover of 0.5% is very small and water bodies are also reduced. To increase the quality of life in an urban environment, it is essential to preserve and manage natural land use classes through appropriate urban planning (Manju et al. 2011).

Assessment of **Raichur** City located in the **state of Karnataka**, **India**, revealed that the city is under imminent threat from the rapid urbanization. As per the present trend, by 2021, nearly 27% of the agricultural land would be converted to settlements resulting in shortage of the surface and ground water resources which, in turn, would further impact agriculture (Basawaraja et al. 2011).

The agricultural space is used, connected agricultural land is destroyed, natural potential of soils is indirectly lost and the endangered animals and plants are eliminated (Ecological Impacts). Inhabitants of densely built cities have to bear lower traffic costs. As households and firms suburbanize, radial commuting to the city centre is more and more replaced by cross-commuting within the urban area. The time cost of commuting would have increased even more without suburbanization (Traffic Impacts). As the number of cars on the road increases, a number of harmful pollutants are emitted as vehicular exhausts. These pollutants include carbon monoxide gas, nitric oxide, hydrocarbons and particulates. Urbanization and anthropogenic stress in **Bangalore City** has paid a share in discontinuity of the drainage network due to loss of wetlands (Zutshi et al. 2008).

The eastward expansion of **Kolkata** due to population growth and influx has been accommodated at the expense of natural ecosystems, mainly the East Kolkata Wetlands. Interestingly, the Basic Development Plan (BDP) for the city, completely disregarding the ecological sensitivity of the EKW, proposes to develop two major townships, namely the Baishnabghata-Patuli Township, and the East Calcutta townships in the EKW. The Sundarbans, which is part of the delta of the Ganga-Brahmaputra-Meghna basin shared between India and Bangladesh, is home to the largest mangrove forest ecosystems in the world, over 1.400 recorded species, including the iconic Bengal Tiger, *Panthera tigris* and several other threatened species (WWF India 2011).

Unfortunately, the natural recovery of the **Navi Mumbai** corridor's mangroves is not enough to offset the overall loss of Mumbai's mangroves. Increased urbanization also presents new threats: most imminently, the new proposed airport development in Mumbai presents a great risk, with the potential to destroy much of this newly re-created habitat. Thus, changes in human resource use can have consequences for ecosystem degradation as well as for restoration (Fig. 7.15).

In India, most studies on urbanisation focus on megacities like Mumbai, Delhi, Chennai, Kolkata and Bangalore, by mostly focusing on issues related to urban economy, politics, and environmental pollution. However, it is the urbanisation of smaller cities and towns, particularly those located in biodiversity rich areas that are of serious concern to conservation (Fig. 7). Yet, there are no studies that enumerate the ground realities of urbanization and its impacts on biodiversity and ecosystem in India (Rajashekariah WWF-India 2011).

Fig. 7.15 Rapid infrastructure development cutting through forest areas negatively impacts biodiversity and ecosystem. *Source* (Rajashekariah 2011)

7.5.6 Bangladesh

Dhaka has gone from being small provincial headquarter to being large connected economic, physical, and environmental features of Bangladesh. The problems that society faces because of the growth of Dhaka and a concentration of large number of human populations is just beginning to be fully recognized as a significant **environmental problem** (Khan 2000).

From recent field visit experience in the Dhaka-Narayanganj-Demra DND project area, they observed that the housing companies have acquired cheap land in flood plains and developed residential colonies there, which are very vulnerable to flooding. Because of the global warming many disasters are raising. As a result each year the city dwellers suffer acute water logging problems during the rainy season. In a report on **Strategic Environment Assessment of Dhaka**, water logging has been identified as one of the major problems of the City (Saha et al. 2012).

Serious problems of environmental degradation in **Sylhet city** stemming from urbanization may be evaluated in the form of land use alterations, inadequate solid waste management, traffic congestion, water logging, drainage congestion, inadequate public water supply, lack of proper sanitation, air pollution, noise pollution, hill cutting, lowering of groundwater level, etc. (Alam et al. 2006).

Table 7.11 Resources, processes and effects of urban ecosystems

Resources	Processes	Effects
Human resources Sunlight Land Water minerals Electricity Fuels Finance Intermediary products Recyclable materials	Manufacture Transportation Construction Migration Population growth Residence/living Community services (education, health etc.)	Negative effects: Pollution—air, water, noise waste generation—garbage, sewage congestion, overcrowding Positive effects: Products, Value-addition, Increased knowledge base/education, Access to better services

Source Srinivas (2003)

7.6 Degradation of Functions and Problems to Use Urban Ecosystem Services

There are complex interactions among social, institutional, economic and environmental variables in urban ecosystems. These interactions adversely affect the functioning of ecosystem and their services to humans diminish. Any disturbance in urban ecosystem resulting from human activities in urban areas eventually affect society and human well-being (Alberti 2005) (Table 7.11).

Important urban patterns include urban form and land use intensity, heterogeneity and connectivity. These drivers give rise to different mechanisms such as land cover change, fragmentation, resource use, emissions, etc. Changes in land cover affect soil quality, primary productivity, biotic diversity, urban runoff, and sedimentation rates. Urban activities also affect population, communities, and ecosystem dynamics by altering availability of water, energy and nutrients (Alberti et al. 2003).

7.7 Major Causes of Ecosystem Degradation Are Given as Under

Human-induced factors: One of the important factors is population pressure. Increased urban population results into higher use of resources including fossil fuel and increased waste generation and environmental pollution. Consequently, there will be degraded, scarce, or absent ecological services in urban areas (Figs. 7.16 and 7.17).

Economic Factors: Economic growth is the main driver of urbanization. Urban development causes conflicts which disturb urban habitat and ecosystem.

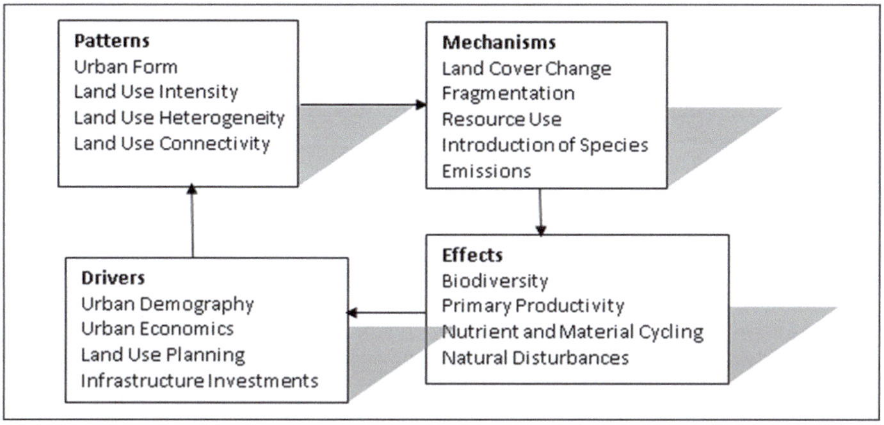

Fig. 7.16 Effects of urban patterns on ecosystem function. *Source* Alberti et al. (2003, designed: M.M. Anwar)

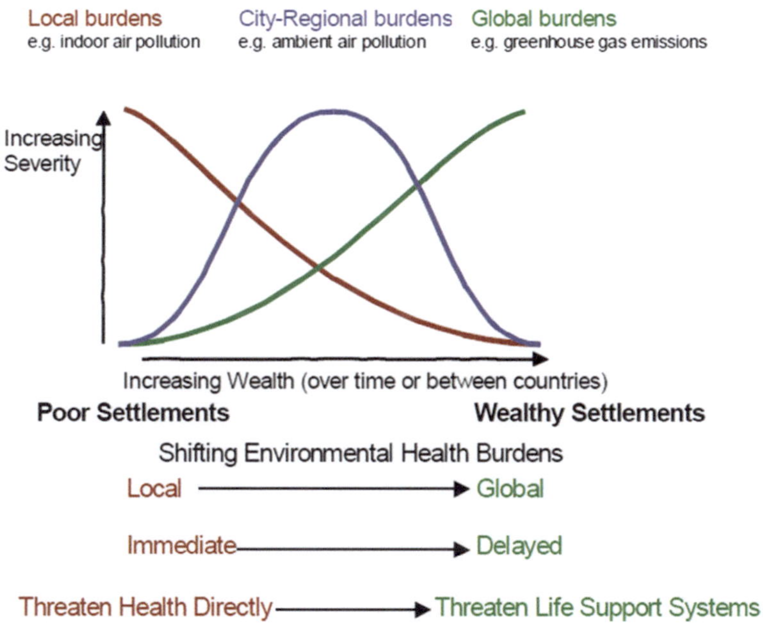

Fig. 7.17 Relationship between economic conditions and urban environments. *Source* McGranahan et al. (2001)

Political Factors: Lack of zoning/growth regulations in urban areas leads to sprawl, conversion of natural areas into residences and loss of open space.

Industrial Projects: There is discharge of wastes from industrial units in urban areas. Industrial development also affects urban ecosystem.

Transportation: Roads and railways cause degradation of ecosystem due to use of fossil fuels, emission of exhaust gases and fragmentation.

Urban development: It involves different activities such as digging, material transport, construction, energy consumption, etc. These activities affect urban ecosystem (Alberti 2005) in following ways;

- Fragmentation and degradation of natural habitats
- Homogenization of species composition
- Disruption of hydrological systems
- Modification of nutrient cycling and energy flow

Urbanization presents major economic challenges, such as high demand for low-income urban employment. There is increased pressure on ability of state to provide basic services such as electricity, housing, healthcare and clean water due to higher migration rate (Kugelman 2013).

There is tremendous population growth and urban sprawl in megacity of Karachi. This has increased travelling distances for urban dwellers and increased use of natural resources (Qureshi 2010). There is reduction in vegetation cover of the natural resources largely due to residential and commercial developments (Qureshi et al. 2010).

Sajjad et al. (2009) found that there is increase in the built-up land from 104 to 200 km^2, and reduction in the overall green cover from 111 to 75 km^2. Urban development has also covered more than 50 km^2 of the open spaces in and around the city of Karachi. Urbanization has resulted to increased temperature up to 4 °C in Karachi during 1961–2009 (Fig. 7.18).

Significant reduction in urban green has been found in urbanized areas of Karachi, particularly after 2000 due to extensive development activities in the

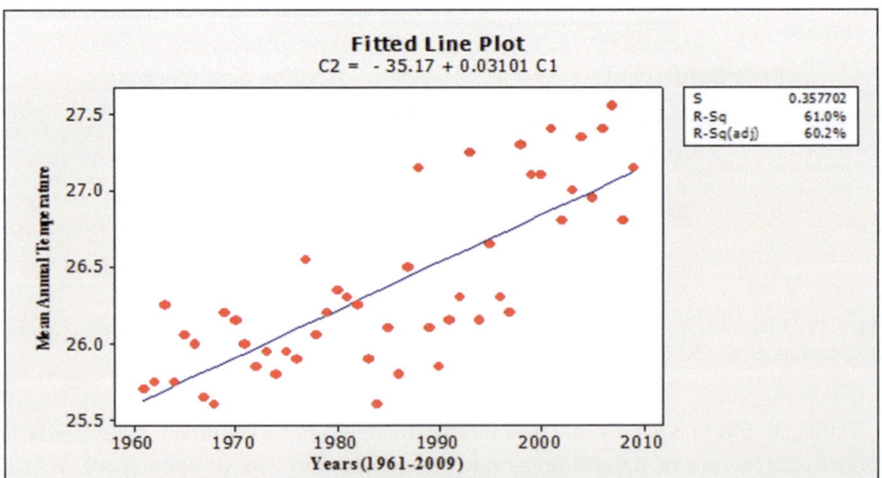

Fig. 7.18 Trend of mean annual temperatures in Karachi (1961–2009). *Source* Sajjad et al. (2009)

transportation sector. There is rapid increasing trends in respiratory and other environmentally induced diseases in the mega city. There are increased physical and social problems in the city due to haphazard and disorganized development activities (Qureshi et al. 2010).

In Dhaka, more than five million people lack access to a public toilet. The public health impact is enormous: In Bangladesh, poor sanitation costs over US$ 3.88 billion a year, which is mostly attributed to healthcare costs. It amounts to 6.3% of the total GDP of the country. About 5.5 million urban poor workers are outdoors in the city for 5–8 h a day and they would likely not have access to a public toilet. There are only 47 public toilets with open access in Dhaka, which means that one-third of population in Dhaka is forced to defecate in open space, along roadsides, or riverbanks (Intellecap 2011).

7.8 Concluding Remarks

Urban development in South Asian is occurring at faster rate to accommodate growing population. This is giving rise to megacities. This is giving rise to many environmental and social issues, which can only be addressed by examining urban a regions as an ecosystem. Thus ecosystem based management approach can provide some practical solutions in new and old urban centres. This chapter summarise some of our current understanding of the regional urban ecosystem and cite examples from South Asia and offer management options for sustainable cities.

Wittgenstein was sometimes criticized for going beyond the aims and methods of Austrian school reform. For example, he successfully taught advanced algebra, geometry, natural history (biology, botany, and geology), and literature. Rosner describes how he 'often went on his own way, wanting to find new methods of teaching and education' (Hausmann 1982:22)[1] Rosner refers not only to Wittgenstein's success at teaching advanced subjects, but also to innovations such as the building and use of a potter's wheel. One of Wittgenstein's pedagogical innovations was to begin with interesting and unusual examples (rather than standard or ordinary ones) not only in mathematics, but also in grammar and other subjects:

> Wittgenstein argued that the child should learn the principle of a thing through an interesting, though possibly difficult specific case; even if other standard examples were easier to learn, there was no point in cluttering up the mind of the child with them unless he

[1]According to Bartley, 'Wittgenstein often poked fun at the reform program's more vulgar slogans and projects' (Bartley 1974:61). It is worth noting that he also made similar comments about philosophical slogans. For example, Redpath writes: 'Wittgenstein intimated that the dictum 'in a great number of cases it is advisable to put "use of a word" for "meaning of a word"' is a slogan. Sometimes the slogan is ridiculed, sometimes it is boosted: in both cases, in his opinion, wrongly…Slogans are easy and stick in the memory. If the use of the slogan goes, but the slogan remains, it is ridiculous' (Redpath 1990:85). Papaneck also notes that Austrian school reform was 'in the best sense of the word, eclectic…[and] creative in its own right' (Papaneck 1962:68).

> understood and could apply the principle behind them. Thus one went from the unusual to the ordinary rather than – as many had hitherto supposed – from the ordinary to the unusual (Bartley 1974:79).

Students learned through investigation, and their active participation involved acquiring knowledge by working or puzzling things out for themselves. Such practical exercises played an important role in Wittgenstein's teaching (Monk 1990:193). In other words, the meaning of specific operations was discovered while their techniques were being acquired (Bartley 1985:98).[2]

References

Alam MJB, Rahman MH, Khan SK, Munna GM (2006) Unplanned urbanization: assessment through calculation of environmental degradation index. Int J Environ Sci Technol 3(2): 119–130

Alberti M (2005) The effects of urban patterns on ecosystem function. Int Reg Sci Rev 28(2): 168–192

Alberti M (2009) Advances in urban ecology: integrating humans and ecological processes in urban ecosystems. Springer, New York

Alberti M, Marzluff J, Shulenberger E, Bradley G, Ryan C, Zumbrunnen C (2003) Integrating humans into ecology: opportunities and challenges for studying urban ecosystems. Bioscience 53(12):1169–1179

Arif GM, Hamid (2009) Urbanization, city growth and quality of life in Pakistan. Eur J Soc Sci 10 (2):196–215

Bajawa IU, Ahmad I, Khan Z (2005) Urban housing development in Pakistan; a case study of Lahore metropolitan area 248:65–73

Bastian O, Syrbe R-U, Rosenberg M, Rahe D, Grunewald K (2013) The five pillar EPPS framework for quantifying, mapping and managing ecosystem services. Ecosyst Serv 4 (2013):15–24

Breuste J (2002) Urban Ecology. In: Bastian O, Steinhardt U (eds) Development and perspectives of landscape ecology. Kluwer Academic Publishers, Dordrecht, pp 405–414

Breuste J (2009) Structural analysis of urban landscape for landscape management in German cities. In: McDonnell M, Hahs A, Breuste J (eds) Ecology of cities and towns: a comparative approach. Cambridge University Press, Cambridge, pp 355–379

Breuste J (2011) Ecology in cities: man-made physical conditions. Introduction. In: Niemelä J, Breuste J, Elmqvist T, Guntenspergen G, James J, McIntryre N (eds) Urban ecology, patterns, processes, and applications. Oxford University Press, Oxford, pp 17–18

Breuste J, Feldmann H, Uhlmann O (eds) (1998) Urban ecology. Springer, Berlin (714 pp)

Breuste J, Haase D, Elmqvist T (2013) Urban landscapes and ecosystem services. Wratten S, Sandhu H, Cullen R, Costanza R (eds) Ecosystem services in agricultural and urban landscapes. Wiley, Chichester, pp 83–104

CIA (2013) The World Factbook. https://www.cia.gov/library/publications/the-world-factbook/fields/2212.html

Costanza R, d'Arge R, de Groot R, Farber S, Grasso M, Hannon B, Limburg K, Naeem S, O'Neill R V, Paruelo J, Raskin R G, Sutton P, van den Belt M (1997) The value of the world's ecosystem services and natural capital. Nature 387:253–260

[2]Bartley quotes this passage from Robert Dottrens's The New Education in Austria (1930) (New York: The John Day Co.) p. 59.

CDGK, Current condition and issues. http://14.192.147.139/cdgk/Home/Departments/tabid/199/Default.aspx. Accessed 21 Feb 2014
Datta P (2006) Urbanisation in India. Population Studies Unit Indian Statistical Institute 1–16
DeGroot R, Wilson MA, Boumans RMJ (2002) A typology for the classification, description and valuation of ecosystem functions, goods and services. Ecol Econ 41:393–408
Demographia World Urban Areas (2013) Urban Agglomerations, 9th Annual Edition
Dutt AK, Noble AG (2004) Urban development in South Asia. In: Dutt AK et al (eds) Challenges to Asian urbanization in the 21st century. Springer, Netherlands, pp 255–275
Elmqvist T, Alfsen C, Colding J (2008) Urban systems. In: Jørgensen SE, Fath BD (eds) Ecosystems, vol 5 of encyclopedia of ecology. Elsevier, Oxford, pp 3665–3672
Endlicher W, Hostert P, Kowarik I, Kulke E, Lossau J, Marzluff J, van der Meer E, Mieg H, Nützman G, Schulz M, Wessoloek G (2011) Perspectives in urban ecology. Springer, Berlin
Gaston KJ (ed) (2010) Urban ecology. Cambridge
Gaston KJ, Davies ZG, Edmonds JL (2010) Urban environments and ecosystem functions. In: Gaston KJ (ed) Urban ecology. Cambridge, pp 35–52
Gilbert OL (1989) The ecology of urban habitats. London/New York
Joshua G, Ali Z (2011) Avian diversity with the varying urban congestions of Lahore. J Ani Plant Sci 2(21):421–428
Khan NI (2000) Temporal mapping and spatial analysis of land transformation due to urbanization and its impact on surface water system: a case from Dhaka metropolitan area, Bangladesh. In: International archives of photogrammetry and remote sensing
Kugelman M (2013) Urbanisation in Pakistan: causes and consequences. Norwegian Peacebuilding Resource Centre (NOREF), pp 1–7
Marten GG (2001) Human ecology: basic concepts for sustainable development. London
Marzluff JM, Shulenberger E, Endlicher W, Alberti M, Bradley G, Ryan C, Zumbrunen C, Simon U (2008) Urban ecology. An international perspective on the interaction between humans and nature. New York
Mathur O (2011) Urban challenges in South & South-West Asia. United Nation ESCAP (Economic and Social Commission for Asia and the Pacific)
McDonald R (2009) Ecosystem service demand and supply along the urban-to-rural gradient. J Conserv Plan 5:1–14
McDonnell M, Hahs A, Breuste J (eds) (2009) Ecology of cities and towns: a comparative approach. Cambridge University Press, Cambridge
McGranahan Gordon, Jacobi Pedro, Songsore Jacob, Surjadi Charles, Kjellén Marianne (2001) The citizens at risk: from urban sanitation to sustainable cities. Earthscan, London, p 17
Millennium Ecosystem Assessment (2005) http://www.millenniumassessment.org/en/. Accessed 14 June 2014
Niemelä J, Saarela S-R, Söderman T, Kopperoinen L, Yli-Pelikonen V, Väre S, Kotze DJ (2010) Using the ecosystem service approach für better planning and conservation of urban green spaces: a Finland case study. Biodivers Conserv 19:3225–3243
Niemelä J, Breuste J, Elmqvist T, Guntenspergen G, James P, McIntyre N (eds) (2011) Urban ecology, patterns, processes, and applications. Oxford University Press, Oxford
Pauleit S, Breuste J (2011) Land use and surface cover as urban ecological indicators. In: Niemelä J, Breuste J, Elmqvist T, Guntenspergen G, James P, McIntyre N (eds) Urban ecology, patterns, processes, and applications. Oxford University Press, Oxford, pp 19–30
Qureshi S (2010) The fast growing megacity Karachi as a frontier of environmental challenges: urbanization and contemporary urbanism issues. J Geogr Reg Plan 3:306–321
Qureshi S, Kazmi SJH, Breuste J (2010) Ecological disturbances due to high cutback in the green infra-structure of Karachi: analyses of public perception about associated health problems. Urban Forest Urban Greening 9:187–198
Rajashekariah K (2011) Impact of urbanisation on biodiversity case study from India. WWF, India
Ravetz J (2000) City region 2020. Integrated planning for a sustainable environment. Earthscan, London
Rück M (2005) Megastädte – Megarisiken. Trends und Herausforderungen für

Saha B, Tarek MH, Saha P (2012) Quantification of the environmental degradation due to the urbanization of Dhaka Narayanganj Demra (Dnd) project area in Bangladesh. Glob J Inc. (USA) 12(01)

Sajjad SH, Hussain B, Khan MA, Raza A, Zaman B, Ahmad I (2009) On rising temperature trends of Karachi in Pakistan. Clim Change 96:539–547

Santos LD, Martins I (2007) Monitoring urban quality of life—the porto experience. Soc Indic Res 80:411–425

Sattar S (2012) Rapid urbanization as a source of social and ecological decay: a case of Multan City, Pakistan. Asian Soc Sci 8:180–189

Schneiders A, van Daele T, van Landuyt W, van Reeth W (2012) Biodiversity and ecosystem services: complementary approaches for ecosystem management? Ecol Ind 21:123–133

Srinivas H (2003) The ecosystems approach to urban environmental management. United Nations Environment Programme. UNEP-IETC

United Nations (2007) World urbanization prospects, the 2007 revision, New York

van Oudenhoven APE, Petz K, Alkemade R, Hein L, de Groot RS (2012) Framework for systematic indicator selection to assess effects of land management on ecosystem services. Ecol Ind 21:110–122

Westerink J, Buizer M, Santiago Ramos J (2002) European lessons for green and blue services in The Netherlands. Working paper for governance and strategic planning scenarios at www.plurel.net

WHO (2016) Global urban ambient air pollution database. WHO_AAP_database_May2016_v3web.xlsx. http://www.who.int/phe/health_topics/outdoorair/databases/cities/en/. Accessed 1 Feb 2017

Versicherung und Risikomanagement. www.munichre.com/publications/302-04270_de.pdf

United Nations (2009) World urbanization prospects, the 2009 revision, New York

Zutshi B, Prasad SGR, Nagaraja R (2008) Anthropogenic impact on the lakes ecosystem in Hi-Tech City, Bangalore, Karnataka, pp 1786–1793

Chapter 8
The Hydro-Institutional Challenge of Managing Tanks: A Study of Tanks in Rural Andhra Pradesh

M. Dinesh Kumar and Neena Rao

8.1 Introduction

Tanks have been an important source of irrigation in India for generations. The states of Andhra Pradesh, Karnataka and Tamil Nadu have the largest concentration of irrigation tanks, numbering 1.2 lac (Palanisami et al. 2010), and accounting for nearly 60% of India's tank-irrigated area (Karthikeyan 2010). They play the vital role of harvesting surface runoff during monsoon and then allowing it to be used later. The predominance of tanks in the Deccan plateau is because of the unique topographic characteristics of the regions. The areas falling under these regions offer ideal potential for tank construction and carrying out gravity based irrigation (ADB 2006).

Tanks are very important from an ecological perspective as they help conserve soil, water and bio-diversity (Balasubramanian and Selvaraj 2003). Like other wetlands, tanks play a significant role in supporting food chains by providing an environment where photosynthesis can occur and recycling of nutrients can take place (Juliano and Simonovic 1999). In addition, tanks also contributes to groundwater recharge, flood control and silt capture (Mosse 1999). For instance, tanks help to lessen the impacts of flooding by absorbing water and reducing the flood discharge. Further, during periods of flooding, they trap suspended solids and nutrient load. Thus, the streams passing through will carry lesser amount of suspended solids and nutrients to the rivers than if they flow directly into them (Bassi et al. 2014).

M.D. Kumar (✉)
Institute for Resource Analysis and Policy, Hyderabad, India
e-mail: dinesh@irapindia.org

N. Rao
SaciWATERs, Hyderabad, India

H. Sandhu (ed.), *Ecosystem Functions and Management*,
DOI 10.1007/978-3-319-53967-6_8

Tanks play a vital role of harvesting surface runoff during monsoon and then allowing it to be used later. Although most of the tanks were essentially constructed for irrigation purpose, they have been used for providing water for domestic and livestock consumption. The tank water body is also used for fisheries, and tank bed as a source of nutrient rich land for cultivation, fodder and soil for brick making. These uses have high value in terms of household income, nutrition and health for the poorest of the poor (Kumar et al. 2013). Over the years, the multiple-use dependence on tanks has only increased. The tank irrigated area has been declining in India over the years. In undivided Andhra Pradesh, figures are more alarming with the net area irrigated by tanks reducing to 4.9 lac ha in 2003–04 from 7.47 lac ha in 1995–96, a decline of 35% (Ministry of Agriculture, Govt. of India).

The impact of decline in tank systems on the rural communities, who have been traditionally dependent on these marvelous socially engineered water harnessing systems, is manifold. This dependence is not only for water for irrigation and domestic use but also for the forestry, fisheries, brick-making, manure and fodder. The neglect of tanks has resulted in farmers receiving insufficient quantities of water from tanks (Palanisami 2006). A study conducted in tank irrigated areas of Tamil Nadu estimated reduction in crop yield and income for tank dependent farmers owing to growth of private well irrigation and deterioration in tank performance (Kajisa et al. 2004). Thus, a well- functioning tank system has a significant bearing on the household income especially for the small farmers who have limited private resources to invest in wells and pump-sets.

Recent attempts to modernize and rejuvenate existing irrigation tanks have focused more on physical rehabilitation with little or no emphasis on understanding of the tank hydrology. Particularly, the way land-use changes in the catchment are affecting tank inflows and siltation rates etc. have been paid least attention by those who are involved in tank rehabilitation programmes. The fact is that intensive crop cultivation, often in the common land through encroachments, and intensive pumping of groundwater in the upper catchments for irrigation are likely to threaten the very sustainability of the tank ecology in many areas. Intensive cultivation will impound a significant share of the catchment runoff; whereas excessive groundwater pumping in hilly catchments can reduce groundwater outflows into the streams, which constitute part of the tank inflows, with resultant impact on the ecological health of the stream.

If it is so, this will have serious implications for tank management programmes. From a physical systems perspective, if performance of tanks as wetland ecosystems is to be sustained or improved, it will be important to influence the land-use decisions and groundwater use in the catchment. From an institutional perspective, the domain of the conventional institutions that are being created to manage the tanks by governments and NGOs alike will have to expand to bring groundwater users and catchment cultivators under its fold. This calls for developing entirely new sets of protocols for tank rehabilitation, including physical strategies for management of tanks as wetland ecosystems and institutional arrangement for ensuring their sustainable performance.

8.2 Dominant Theories on Decline of Tanks

A large volume of literature exists on "decline" of tanks in South India. This "decline" is both in terms of declining relative contribution of tanks in irrigated area, and reduction in aggregate area under tank irrigation (Sharma 2003). However, the theory of "decline of tanks" is contested by a few scholars, who believe that this emerges from a very reductionist approach of viewing tanks as mere sources of irrigation, and that the criteria for evaluating the performance of tanks should be more broad, to accommodate their various social, economic and ecological functions (Kumar 2002; Palanisami et al. 2010). Kumar (2002) for instance, argue that given the wide range of physical, social, economic and ecological functions which tanks perform, the simplistic criteria for evaluating the performance of tanks, which look at the irrigated area or the number of users of drinking water as the indicators, need to change and more complex criteria need to be evolved.

Nevertheless, various theories have been made as reason for the "decline of tanks" in South India. Shankari (1991) points out that poor management is primarily responsible for the decline of tanks, as evident in the nonparticipation of farmers in cleaning channels, encroachment of the tank bed, inadequate repairs, weed infestation and siltation (Shankari 1991). von Oppen and Rao (1987), based on a survey of 32 tanks in Andhra Pradesh and Maharashtra, argued that increases in population density resulted in deforestation in catchment areas leading to soil erosion and siltation. After, Sekar and Palanisami (2000), tank bed cultivation and the lack of an administrative structure to provide timely repair and maintenance, contributed to the decline of tank irrigation. The other reasons provided by scholars include: agricultural encroachment of supply channels and tank beds, which reduced the inflows into the tanks (Easter and Palanisami 1986; Mosse 1999); sand mining of supply channels; rural infrastructure development interfering with the natural inflows; and unplanned watershed development cutting off the supply to tanks (ADB 2006; Palanisami 2006); decline in tank storage capacity over the years due to excessive siltation (Gunnell and Krishnamurthy 2003; Paranjape et al. 2008); lackadaisical attitude of micro-level institutions managing the tanks, which has mainly stemmed from the growth of private well irrigation in the tank command area resulting in disincentive among farmers to manage these open access bodies (Balasubramanian and Selvaraj 2003; Kajisa et al. 2004; Sakthivadivel et al. 2004; ADB 2006). A few scholars have also highlighted how the hydraulic interdependence between tank storage and aquifer recharge is creating disincentive for farmers to carry out maintenance of tanks, and instead motivating them to privatize these resources by using them as percolation ponds (Sakurai and Palanisami 2001).

A multi-variate analysis by Balasubramanian and Selvaraj (2003) of the factors responsible for tank degradation showed that variables such as encroachments in catchment and water spread area and the increase in canal- and well-irrigation, had significantly increased the degradation of tanks. The increasing importance given to modern irrigation systems, larger reservoirs and river valley projects and the spread of private irrigation wells, also have a considerable negative impact on traditional community irrigation systems.

A common view which emerges from the review is that the lack of interest of command area farmers in the management of tanks (Shankari 1991; Balasubramaniyam and Selvaraj 2003; Sekar and Palanisami 2000; von Oppen and Rao 1987), or the erosion of the community management structures, which were responsible for their management, and the subsequent management take over by the government, had resulted in their decline (PRADAN 1996; Rao 1998). Some attributed the loss of community interest in tanks to the advent of groundwater irrigation which gave farmers superior control over irrigation (Dhawan 1985; Palanisami and Easter 1991). Some scholars attribute it to the development approach followed during the British rule centered on modern large irrigation systems for the decline of tanks (Paranjape et al. 2008).

However, Mosse (1999) challenged the long held view of scholars working on tanks in South India that collapse of community institutions was the major cause of decline of tanks. He contends that even in the past, communities did little investment in the upkeep of tanks. It was the Zamindars and kings who, not only built most of the tanks, but spent money for their upkeep as well. According to Mosse (ibid), it was the fall of the institution of overlords that led to the decline of the tanks.

Shah (2008), while highlighting the cultural and environmental superiority of traditional knowledge that built and managed tanks, as an irrigation technology, argued that tanks have not necessarily produced a democratic social order, either in the past or in the present. According to her, they symbolized an increasingly extractive statecraft involving coerced labor; the expropriation of surplus by elites; and the spread of technological choices that could be environmentally unsound and that often resulted in forced displacement, uncertainty, technological vulnerability, and social anxiety and violence. Further, tanks as techno-sociological artifacts were socially embedded in societies and economies that were organized for warfare, and that sustained sharp social hierarchies, and were often violent to women and people from lower castes (Shah 2008: p 673).

In nutshell, when viewed together, the work of various scholars seem to suggest that decline of tanks was due to several events, starting with the takeover of community and Zamindari (Private) tanks by the state, which led to institutional erosion, collapse of the system for collection of water charges and lack of maintenance leading to deterioration of the physical condition of this irrigation infrastructure, and the subsequent gradual loss of community interest in their affairs. Therefore, the factors responsible for decline of tanks were argued to be "institutional".

None of the above theories could fully explain the reasons for degradation of tank systems and decline in tank irrigated area in South India. In fact, every argument suffers from weaknesses. For instance, if the argument that advent of groundwater irrigation had really led to the loss of farmer interest in tanks, is valid, then it tends to assume that groundwater irrigation is highly equitable and provides farmers from all segments access to and control over well irrigation. This is far from the truth, as pointed out by Kumar (2007). Only a small fraction of the small and marginal farmers in India even today own wells, and pump sets (Kumar 2007), where the situation was much worse in the 70s and 80s, when drilling wells was expensive, and rural electrification was poor.

As pointed out by Narayanamoorthy (2007), it is the small and marginal farmers, who do not own irrigation wells in the tank commands, who have high stakes in tank irrigation as their livelihood is heavily dependent on it. If this is the case, one cannot explain the poor state of affairs with regard to the condition of tanks in areas such as Kolar (in Karnataka) and Anantapur in AP, where the poor small and marginal farmers do not own wells, and are often dependent on water purchase from well owners engaged in water trading. It is also a notable fact that most of the beneficiaries of tank water for irrigation are small and marginal farmers.

8.3 An Alternative Hypothesis on Tank Degradation

The dominant theories, advanced by many researchers, harp on collapse of traditional tank institutions as the cause of decline of tank irrigation, be it the institutions of overloads, or community management structures. Both the theories are also suggestive of a resounding view that performance of tank as a system is very much within the control of the farmers in the command areas or the institutions which manage them. They inadvertently ignore the fact that these institutions existed in a certain socio-political landscape, which is difficult or even impossible to recreate. Such views are based on the assumption that simply cleaning supply channels, or clearing the catchments or repair of the tank embankments (tank bunds), and de-silting of the distribution network would yield results in terms of improved storage in tanks and expanded irrigation benefits. They ignore the effect of endogenous and exogenous physical and socio-economic factors on tank hydrology, and the very impact these effects have on the viability of institutions itself.

First: population growth had a significant demand on irrigation water for crop production in these semi-arid regions, which forced farmers to go for alternative sources of irrigation, and the search for this alternative was facilitated by advent of well irrigation, cheap drilling technology, rural electrification and subsidized electricity for pumping groundwater. This is quite understandable given the fact the performance of tanks was subject to high variability in accordance with year to year variation in the occurrence of monsoon rains, as these tanks harnessed water only from the local catchments. Further, with a manifold increase in rural population and with increase in number of farmers within the limited command of these tanks, the actual area which a single farmer could irrigate using tank water became too low for them to manage their farming enterprise. Well irrigation not only became affordable and in some cases cheaper, it also provided a superior form of irrigation. Yet, as noted by Narayanamoorthy (2007) and Paranjape et al. (2008), for the poor small and marginal farmers, the tanks continued to be an important source of irrigation and livelihoods.

Second: the increase in population pressure on private land also meant that farmers had to expand the net area under cultivation, and sometimes this led to encroachment of commons, which formed the original catchments of tanks. Catchment cultivation resulted in a lot of the runoff generated from precipitation being captured through in situ water harvesting for production of rain-fed crops, reducing the inflows into the tanks. Intensive well irrigation on the other hand led to reduced groundwater outflows (base flow) into the upper catchment tanks. Whereas draw down in water table resulting from excessive withdrawal of groundwater can potentially lead to greater percolation of water from tanks into the shallow aquifers, further reducing the storage and irrigation potential of the tanks. It is important to mention here that as research has shown in the past, in hard rock areas of peninsular India, the "cone of impression" produced at the bottom of tanks due to percolation of tank water into the shallow aquifer generally stops further percolation of the water (Muralidharan and Athavale 1998). This peculiar geo-logical and geo-hydrological setting ensured storage of water in the tanks in peninsular India. But, emptying of the aquifer could induce sustained recharge. What is important is that there is no mechanism to control groundwater abstraction by well irrigators in the command and catchments, as legal rights to groundwater are attached to landownership rights.

Reduced irrigation potential of tanks due to the above cited reasons and the increasing number of tank water users in the command area essentially meant that the contribution tanks could make in the overall livelihood of individual command area farmers, including small and marginal farmers, was too small in comparison to the transaction cost of initiating actions that would improve their performance. In fact, the transaction cost of initiating actions such as removal of encroachments from catchments, and regulating the use of groundwater in the command and catchment would be too high owing to the complicated legal formalities involved, whereas what a farmer could earn in terms of income from crop outputs that can be produced from the use of tank water could be quite insignificant. Whereas, many of the other engineering interventions such as stabilization of tank bund, increasing the capacity of tank through de-silting and clearing of tank catchments do not result in incremental benefits that commensurate with the financial investment in most situations. This significantly reduced the incentive among the members of the farming community for self-initiated management actions.

While there are a few tank management activities which can provide substantial private benefits (to the farmers) such as de-silting, i.e., removal of silt and clay from the tank bed, as the use of it in the field gives direct income benefit to them in terms of higher crop yields for two to three years consecutively (Kumar et al. 2011; Paranjape et al. 2008), often the communities lack the wherewithal to take it up. This also explains why the communities come forward to take up tank management activities in situations where there is external support to cover the transaction cost, and the cost of undertaking physical activities.

While the tank potential for meeting the demands for economic activities has been gradually declining in many areas due to the problems described above, improved availability of good quality water from public water supply schemes within the close vicinity of their dwellings has, to a great extent, reduced the dependence of village communities on traditional sources of water such as tanks and ponds for domestic water supplies. Instead, they now they now depend on these tanks for livestock drinking and washing. This has reduced the village communities' incentive and motivation to protect tank water quality for ensuring potability. Simultaneously, the tanks and ponds have become the natural sink for agricultural runoff containing fertilizer and pesticide residues from upper catchments. This is particularly the case for cascade tanks, wherein the upper catchments of some tanks consist of the command area of the tanks located upstream. Presence of nitrates in agricultural runoff had caused eutrophication of tanks and ponds, more so in the case of high rainfall areas, affecting fish populations in these wetlands.

While it is extremely difficult to say which process has resulted in what outcome, the growing awareness of public health impacts of poor water quality, and recognition that traditional water bodies have increasingly become polluted had also forced the government to think about formal water supply for villages. Here again, it is not just that the communities have lost interest in protection of tank water quality. There could be situations where there are no alternative sources of water supply in the village. Even in such cases, when the communities desire to protect the tanks from pollution, there is hardly anything which it can do to stop it. Here, both the polluters and those who suffer the damage caused by pollution are in most cases the same. But, there is a lack of institutional capability to address this issue.

All these factors might have ultimately led to tank degradation in terms of water availability and water quality wherever it has happened. Or in other words, attributing the decline of tanks to one causal factor, i.e., erosion of community institutions or collapse of traditional village institution of "Zamindari" system would be mere over-simplification of a complex physical, socio-economic, cultural, environmental and institutional change. Again, what has really led to decline of tanks in a particular situation is very much case-specific, depending on how the physical, socio-economic, cultural, environmental, and institutions factors have played out in that particular case. But, in sum, we would like make the proposition that technological, socio-economic, institutional and cultural changes happening in the societies within which the tanks are embedded, have resulted in major physical (hydrological and environmental) changes in the tank ecology, which had changed the incentive structures for tank management.

The schemes of minor irrigation departments, NGOs and donor communities in rehabilitation of tanks have their accent on engineering works viz., bund stabilization, construction of weir and repair of sluices, catchment forest clearance and channel de-silting, and institutional development, comprising formation of water user associations and their training for capacity building. They do not pay much attention to the hydrological characteristics of tanks, which are chosen for rehabilitation. Hence they produce poor results, in terms of improving the overall tank performance.

If changes happening in the rural society on the technological, socio-economic, cultural and institutional front are bringing about irreversible trends in tank ecology through changes in hydrology and water environment, then programmes and projects to improve the condition and performance of tanks that are based on simple engineering interventions such as de-silting, catchment clearing (removal of trees), and supply and distribution channel cleaning (silt removal) are unlikely to lead to any beneficial outcomes. Or in other words, if the rehabilitation programme has to be successful, there should be sufficient incentive among the potential tank users to take up the rehabilitation work.

8.4 Research Objectives, Approach and Methodology

The specific objectives of the study were as follows: (1) analyze the impact of well development and groundwater intensive use in the tank catchment and commands on tank performance in terms of irrigated area; (2) analyze the impact of catchment cultivation practices on tank performance in terms of area irrigated; (3) identify the physical, socio-economic, institutional and environmental factors that result in good overall performance of tanks; and, (4) evolve the criteria for selection of tanks for rehabilitation, and work out the broad management strategies for sustaining and improving the performance of the selected tanks.

The study involved an eclectic approach. It used analyses of primary data collected from selected performing and non-performing tanks, which were both quantitative and qualitative in nature (for six selected systems) along with secondary data collected from the state minor irrigation department on tank and well irrigation and land use in catchments (on a time scale) at different scales (state, district, block and individual tanks) for addressing the key research questions. The tanks were selected in such a way that the hydrological and socio-economic environments are not uniform.

8.5 Case Study Regions and Data Sources

A total of six tanks from three districts of the state of Andhra Pradesh were chosen for the study, with two tanks from each district. The locations were chosen in such a way that each one represent a unique situation vis-à-vis the changes in historical performance of tanks. The districts are Vijayanagaram, Nizamabad, and Kurnool. Each one falls in a different river basin. Vijayanagaram falls in the drainage area of one of the east flowing rivers, north of Godavari river basin, which is a water rich river basin. Nizamabad falls in the drainage area of Godavari river basin, which again is a water-rich river basin. The area of Kurnool, which was chosen for the tank case studies, falls in the drainage area of Krishna river basin, which is a water-scarce river basin. Hence, each one represents one hydrological regime.

In Nizamabad district, as the district level data show, the area under tank irrigation, in terms of net tank irrigated area, had undergone dramatic reduction, over the past three and a half decades. In Vijayanagaram district, the area under tank irrigation had undergone some declining trend during the same period, but not as much as that of Nizamabad. In Kurnool district, no major reduction in area under tank irrigation was observed from the time series data for 35 years from 1970–71 to 2004–05.

Within each district, the two study tanks were selected in such a way that one is in a better condition as compared to the other. The judgment about relative performance was made on the basis of the discussions with the officials of the irrigation and CAD department of the government of Andhra Pradesh, who were concerned with the management of these tanks. The criteria used for judging the performance are the condition of the tanks vis-à-vis the physical infrastructure, the area irrigated in the command area against the design command and the community involvement in their management.

Data were collected from both primary and secondary sources. The primary data were collected both at the tank level and the farmer level. The tank level data included: the tank command area; the different uses of tanks and the number of HHs depending on the tank for various uses; area under different crops in different seasons and at different points of time (1970, 1980, 1990, 2000 and at present), and irrigated area under different crops in different seasons and at different points of time; number of wells in the tank command at different points of time; number of wells in tank catchment at different points of time; area under different crops in different seasons in the tank catchment, and irrigated area under different crops in different seasons at different points of time in the tank catchment. They were obtained from the village elders and tank water user association office bearers. The farmer level data consist of: (i) area under different crops in different seasons, and at different points of time; (ii) irrigated area under different crops in different seasons and at different points of time; (iii) changes in cropping pattern and irrigated cropping pattern in drought year, normal year and wet year; and (iv) the current sources of irrigation.

The secondary data comprises the net area irrigated by tank over the period from 1970–71 to 2005–06 in all the districts of AP; the net area irrigated by wells and bore wells over the same time period in all the districts of AP; the area irrigated by different sources viz., wells, tanks and canals in the state over the period 1970–71 to 2005–06; and the characteristics of tanks in different districts vis-à-vis their total water spread area, and the net area irrigated.

8.6 Results and Discussion

8.6.1 Characteristics of Tanks in Erstwhile Andhra Pradesh

Tanks are manmade wetlands as per wetland classification. Out of the total of 18,701 manmade wetlands in the state of (undivided) Andhra Pradesh, with wetland area less than 2.5 ha and covering a total wetland area of 610,354 ha, 15,290 are tanks/ponds constituting a total wetland area of 1.4 lac ha. The rest are man-made reservoirs, water logged areas etc. The state has three distinct physiographical units. First is the coastal plain to the east extending from the Bay of Bengal to the mountain ranges; Eastern Ghats which form the flank of the coastal plains; and plateau to the west of the Eastern Ghats. Maximum tanks are found in the plateau. The district-wise number of tanks, their wetland area, and the water-spread area pre and post monsoon are given in Table 8.1.

What is most interesting is the fact the water spread area of these wetlands shrinks drastically after the winter, touching the lowest point during peak summer. Table 8.1 shows that summer water spread area (73,749 ha) is less than half of post monsoon, i.e., November, 2010 (1,632,277 ha). This has major implications for the total water availability of these tanks and the various functions that these tanks can perform in different seasons. Comparison of district-wise data shows that Vijayanagaram has the largest number of tanks, followed by Nellore and Medak. But, in terms of average size, tanks/ponds in Cuddappa are the largest, with an area of 25.40 ha.

Another interesting observation is that the ratio of the area irrigated by the tank and the wetland area, which reflect the physical characteristic of tanks, vary widely between districts. For the analysis, we have considered the area irrigated by the tanks in 1970–71, assuming that the real deterioration in tank performance started only later. The wetland area (as estimated through remote sensing imageries) of the tank was taken from the Wetland atlas prepared by the Indian Space Research Organization, Ahmedabad. Since the estimates of wetland area do not consider the tanks with wetland area less than 2.5 ha, it might induce some errors in the estimation of irrigated area-wetland area ratio. The ratio varies from 2.62 in the case of Medak to 23.47 in the case of Srikakulam. Here, Gunter was not considered for the analysis owing to the fact that the tanks in this coastal district receive water from large irrigation schemes. Hyderabad was also not considered as data on tank irrigation in the district was incomplete.

8.6.2 Tank Management Programme in Erstwhile Andhra Pradesh

The Irrigation and CAD department of the Government of Andhra Pradesh had undertaken an ambitious programme for rehabilitation of tanks coming under the minor irrigation department, which have design command area of more than 100 acres. It is to be mentioned here that those tanks which have less than 100 ha of design command area are under the jurisdiction of Panchayats. The programme funded by the World Bank, called Andhra Pradesh community based tank management project, envisages rehabilitation of 3000 minor irrigation tanks covering 21 districts of the state. The technical or engineering interventions under the

Table 8.1 Number of tanks and their wetland area in different districts of Andhra Pradesh

Sr. No.	Name of district	Total number of tanks/ponds	Total wetland area (ha)	Average wetland area per tank/pond (ha)	Total water spread area (ha)	
					Post-monsoon	Pre monsoon
1	Adilabad	590	7383	12.51	5090	1947
2	Nizamabad	915	19,152	12.51	13,581	7629
3	Karimnagar	692	10,535	20.93	8802	3590
4	Medak	1066	20,116	15.22	16,250	8011
5	Hyderabad	12	107	18.87	71	73
6	Rangareddy	281	3287	8.92	2728	2002
7	Mahbubnagar	340	5424	11.70	3509	2510
8	Nalgonda	601	11,702	15.95	8573	5857
9	Warangal	659	9462	19.47	8433	2367
10	Khammam	558	8044	14.36	5547	2762
11	Srikakulam	460	4678	14.42	3260	1459
12	Vijayanagaram	1539	10,567	10.17	7724	2293
13	Visakhapatnam	543	4793	6.87	3409	1341
14	East Godavari	505	4244	8.83	3384	2764
15	West Godavari	334	3191	8.40	2409	1808
16	Krishna	511	6312	9.55	4452	2418
17	Guntur	377	3512	12.35	2678	1939
18	Prakasam	480	7490	9.32	5918	3572
19	Nellore	1088	27,673	15.60	24,890	7151
20	Cuddappa	220	3584	25.43	3069	1458
21	Kurnool	142	1887	16.29	1395	810
22	Anantapur	447	9896	13.29	3131	4135
23	Chittoor	1391	18,638	22.14	17,200	3560
24	Total	15,290	212,244	13.88	163,227	73,749

Source Authors' own estimates based on National Wetland Atlas, Andhra Pradesh prepared by Indian Space Research Organization, Ahmedabad

rehabilitation programme include de-silting, jungle clearance, stabilization of bund, sluice repair, waste-weir construction or repair.

As per the project guidelines, the planning, implementation and post construction operation and management of the project are to lie with the water users' associations. Software inputs include training of water users associations for capacity building. To facilitate the planning, implementation and post implementation management of various tanks associated with tank rehabilitation, there would be one support organization at a level of a cluster of 5–10 tanks.

One of the positive features of the programme is that it is demand driven. The type of activities to be undertaken for rehabilitation of tanks for restoration of irrigation in the command area is to be decided by the water users' association. The funds allocated for this, i.e., Rs. 25,000 per ha of design command area, appears to be quite sufficient at current prices, given the fact that this is for retrofitting of an already built irrigation infrastructure, and not for building new ones. Again, the programme has inbuilt checks and balances for technical soundness and financial management. The plans prepared by the Water Users' Associations are scrutinized by the technical staff of the minor irrigation department after field visits, and they prepare detailed project report with the estimates for the planned works subsequently. There is a technical manual, which provides the guidelines for design and execution of the physical works, and system operation.[1]

As the department slightly deviates from its usual operations for this project and added several new activities as the project components, a separate financial manual was prepared exclusively for implementing this project. This document describes the procedure for fund flow and arrangement at the state, district and WUA level, budget preparation, financial management systems at various levels. The manual also provides procedure for fund disbursement and accounting system that will be followed in the project. One important financial innovation used in the project is that any expenditure for rehabilitation work below Rs. 50,000 could be sanctioned by an officer in the rank of Assistant Engineer, and any expenditure in the range of Rs. 50,000–Rs. 200,000 could be sanctioned by a Deputy Engineer or Deputy Executive Engineer.

Another notable innovation is with regard to the repayment of water charges being collected from the tank users. Though the cess is collected by the revenue department office at the Mandal level, arrangements are made for quick release of a major share of these funds back to the Water Users' Association through the concerned section office of the irrigation department. This keeps big incentive for the WUA to motivate the member farmers to clear their water dues timely. These funds are to be used for small repair works.

It appears, from a careful scrutiny of the documents relating to the tank management project, that a clear cut protocol for picking up tanks for rehabilitation is

[1]The manual clearly specifies the guidelines on engineering design, construction and quality control for the physical components of the work. It also clearly specifies the standards that need to be maintained for each of the tank constituent be it concrete work, earth work, reinforcement, stone masonry work, revetments, protection of upstream and downstream works and canal lining.

absent. There was no clear cut policy on the part of the department, which could have been used as the guideline for short-listing the 3000 tanks for rehabilitation, among the many thousands of tanks/ponds in the state. The proposal for the tank management project was prepared on the basis of the feedback and suggestions given by the technical officers (engineers) of the department who are concerned with management of minor irrigation tanks, from the districts and mandals concerned. One quantitative criterion used by the department was to take only those which have a command area of more than 25 ha (i.e., 100 acres). The second criterion was presence of some local initiative by the community for tank management. This, however, was purely based on perception. While this shortcoming could have been overcome through a prudent attempt by the department to involve the local tank communities in this key decision making, there was no involvement of the tank community whatsoever in selecting the tanks for rehabilitation.

Perhaps, one important consideration which was totally missing in the entire selection process is the hydrological condition of the tanks. Very little attention is paid to the fact that several external factors such as the intensive use of groundwater, catchment encroachment for cultivation are ruining the potential of tanks as reliable sources of water for multiple needs, including irrigation, through their effect on hydrology, which is mostly irreversible.[2] Sufficient efforts are not being made to ensure that the tanks chosen for rehabilitation get sufficient inflows from its catchment. This is also evident from the procedure followed for estimating the catchment runoff, one important variable for deciding on the capacity of the tank. The agency appears to be using the "rational formula" for estimating the catchment runoff on the basis of the total catchment area and a runoff coefficient fixed on the basis of the dependable rainfall (for 75% dependability). This is a very crude method of runoff estimation. The problem is that for semi-arid regions with low to medium rainfalls occurring in an erratic fashion, and for the land cover that exist, the runoff will not be a linear function of the annual rainfall magnitude, but will increase with rainfall in an exponential fashion. Hence, assuming a fixed value for runoff coefficient will lead to a major error in estimation of runoff.

Ideally, the runoff coefficient will be very low for low rainfall and disproportionately high for high rainfall in semi-arid regions with high aridity (Prinz 2002) causing high variability in runoff (Kumar et al. 2006, 2008). Therefore, the runoff needs to be estimated on the basis of the rainfall-runoff relationship established for the basin or the sub-basin in which the tank catchment falls. This can be done using historical data on stream-flows and precipitation for the basin/sub-basin under consideration. Once the rainfall-runoff relation (model) is established, the stream flows for the years (for which records are not available) can be computed on the basis of observed rainfall. From these stream-flow values (say for 30–50 years), the dependable runoff (for 75% dependability) can be estimated and used for planning

[2]For instance, it is impossible to stop groundwater exploitation in the catchments and commands. In rural contexts, it is also impractical to clear the encroachments from the catchments.

the rehabilitation work. This has to be the most crucial input for deciding on the types of uses from the tank and the cropping pattern.

In view of the many changes happening in rural landscape including the drainage interception, there is a need for re-assessing the catchment area of the tanks. In the absence of this, there is a very high chance that the catchment and therefore the runoff get over-estimated. The current hydrological planning is flawed, often leading to over-estimation of the tank size, and excavation work.

8.6.3 Impact of Groundwater Irrigation on Tank Inflows

It was hypothesized in the beginning of our research that increase in groundwater use in the catchment and command of the tank would affect the performance of the tanks, through reduction in the inflows into the tank in the form of groundwater outflows or base flows in the streams.

Naturally, base flow contribution to tank inflows would be significant in the case of tanks situated in the upper catchments of river basins, by virtue of the hilly topography and the forested catchments. One example is the tanks located just downstream of the forested catchments of Krishna river basin in Kurnool district. These upper catchment tanks are surrounded on three sides by protected forest area, with good base flows coming in from them. Though in such areas the chances of intensive groundwater use is quite low due to the poor availability of arable land and extremely low groundwater potential in the hard rock formations in the hills, even a small increase in groundwater pumping could alter the hydrological balance.

On the contrary, in the lower catchment tanks, the contribution of base flows to tank inflows could be quite low, and the major contribution of the tank inflows would be from the surface runoff. In these areas, the change in land use is the most important factor altering the tank hydrology. The net area under cultivation in these areas is likely to have increased due to increased pressure on land. Here, in this case, it is not an expansion in area under irrigation which is causing changes in the tank hydrology, but increase in area under cultivation. Mostly, it is resulting from encroachment of the common land, which forms the tank catchment, by villagers for cultivation of rainy season crops, which is causing the change.

While normally a part of the rainwater falling in the natural catchments would infiltrate into the soil and the remaining water would run-off from the land. While part of this infiltrating water would remain in the soil profile depending on the soil storage capacity, the excess water would percolate down the soil strata and join the groundwater table. The water in the soil profile would eventually evaporate if the soil is barren or else it would support the growth of some natural grass species. But, in any case the rate of evaporation of moisture from the soil profile would be very low. Whereas, once the land is covered by crops, the rate of depletion of moisture from the soil profile would be faster, as the crop would take the water for meeting the transpiration needs. While this would create more storage space in the soil profile for the incoming precipitation, the presence of vegetation would increase the

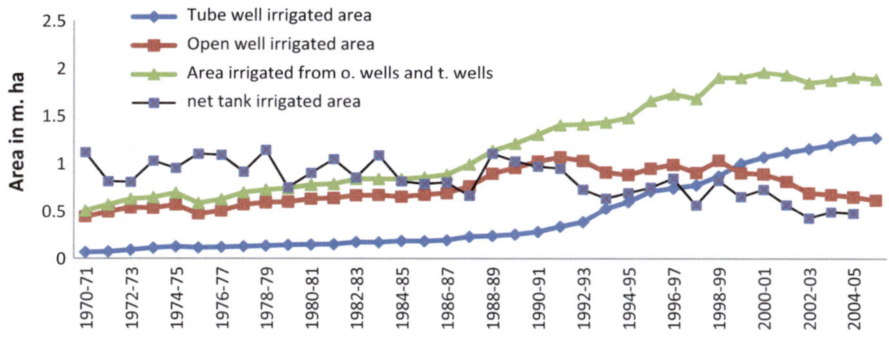

Fig. 8.1 Net area irrigated by open wells, tube wells and tanks, AP

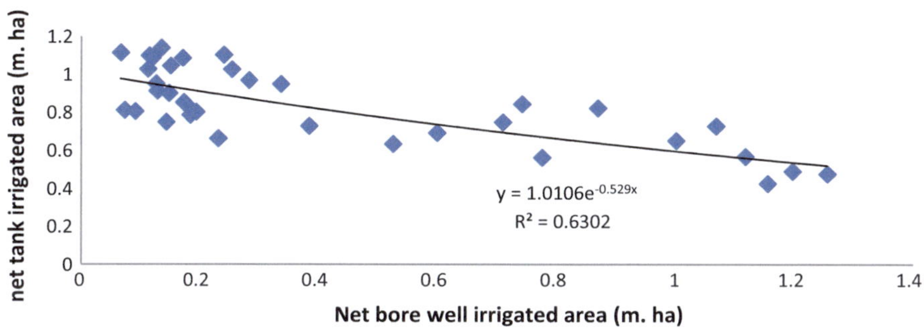

Fig. 8.2 Impact of groundwater irrigation on tank performance

rate of infiltration. The presence of field bunds would further reduce the downward movement of runoff generated in the field, and the water would percolate down the soil if it is in excess of the moisture deficit in the crop root zone.

In order to understand the dynamics of interaction between tank catchment, groundwater and tank hydrology, we began with the analysis of time series data on net area irrigated by tanks, net area irrigated by open wells and tube (or bore) wells. Historical data on net area irrigated by tanks, open wells and tube wells/bore wells for the period from 1970–71 to 2004–05 are analyzed. The results showing the historical changes in the irrigated area from these three sources are presented in graphical form in Fig. 8.1.

Figure 8.2 shows that while the area irrigated by open wells increased till the mid-eighties and touched the peak in 1990–91, and started declining thereafter. Whereas the area under tube wells/bore wells, which tapped the deeper aquifers and geological formations, started increasing exponentially by early 90s and continues even today. Overall, the area irrigated by wells has been increasing till 2000–01, when it peaked at 1.95 million ha.

On the other hand, the (net) area irrigated by tanks started consistently showing declining trends after the late 80s. Though there has been wide fluctuations in the net area irrigated by tanks between years during the previous years, i.e., 1970–71 to 1987–88, such fluctuations could be attributed to inter-annual variability in rainfall, which will have direct impact on tank inflows. Regressions run between net tank-irrigated area and net well irrigated area—the sum of area irrigated by open wells and tube wells—showed a strong inverse relation the two, With increased in well irrigated area, the area irrigated by tanks reduced linearly (R^2 = 0.49). Further, regression was run with net tank irrigated area against area irrigated by bore wells. This showed a sharper and stronger relationship (Fig. 8.2). Going by the regression formulae, a unit increase in net tube/bore well irrigated area resulted in greater reduction in tank irrigated areas, as compared to that caused by unit increase in net total well irrigated area. Also, the increase in tube well/bore well irrigation explained reduction in tank irrigated area to an extent of 63%, against 49% in the case of total well irrigated area (Kumar et al. 2011).

The differential trends can be explained in the following way. Increase in well irrigation resulting from increased withdrawal of water from open wells and bore wells, suggests greater withdrawal of groundwater. This would normally affect the base flows into tanks, and also the percolation of stored water in the tanks into the formations underlying it, depending on where the tanks are located within the basin. But, at the same time, consistent increase in area under well irrigation till around the year, 1991–92 can also be suggestive of the fact that un-sustainable levels of abstraction of the shallow aquifer were not reached till that point of time. Under such circumstances, the effect of groundwater withdrawal on tank inflows will be less. This explains the milder slope of the curve representing decline in tank irrigation with increase in well irrigation.

Now as regards the lower regression value, other than well irrigation, rainfall can also be an important factor which explains the changes in tank irrigated area. In good rainfall years, the tank inflows, including that from groundwater outflows or base flows, could also be high, especially when the open well irrigation is still dominant. Nevertheless, there would be rise in pumping above the normal year values as a result of better replenishment of groundwater. In contrast, in low rainfall years, the runoff and base flows could reduce, along with groundwater recharge. So, in such years, the groundwater irrigation from open wells along with tank irrigation would be less than that of normal year values. This would upset the normal trend. Such trends in well irrigation, which is characteristic of hard rock regions with poor static groundwater resources, are visible in Andhra Pradesh (see Fig. 8.1).

At the same time, increase in tube/bore well irrigation (from 0.98 m. ha in 1990–91 to 1.27 m. ha in 2004–05 by around 0.99 m. ha), which is also accompanied by reduction in open well irrigation, indicates the dewatering of shallow aquifers, and the pumping of water from the deeper strata. This not only means the chances for base flow contributing to streams flows into tanks are almost absent, but the possibility of tanks losing their storage into the dewatered aquifers would also be very high. Again under such geo-hydrological environments of deep dewatered zones, the response of shallow and deep aquifers to incident rainfall would be slower as compared to a situation where the shallow aquifer is saturated owing to the time taken for water to move from the top soil to the water table and the amount of water lost while percolating. Here, unlike in the earlier case, good rainfall may not result in proportional improvement in groundwater recharge and base flows, while tank inflows and tank irrigated area could increase due to increase in runoff. Therefore, rainfall will have a lesser influence on groundwater-tank interactions. This increase the negative effect of tube well/bore well irrigation on tank performance.

While intensive well irrigation explains the decline in tank irrigation to the extent of 48–63%, clearly there are other factors which cause decline in tank irrigated area. We have hypothesized that change in land use would be another important factor, which cause reduction in tank inflows and tank-irrigated area. However, we can test this hypothesis only using primary data collected from the field, as the secondary data on land use are not available for individual tank catchments, but also mandals and blocks and districts.

8.7 Differential Impacts of Groundwater Irrigation on Tank Irrigated Area Across Different Regions

The foregoing analyses do not, however, suggest that the historical tank performance has been uniform across the board. In fact, analysis of tank irrigated area, well irrigated area and bore well irrigated area for different districts show distinctly different trends. For instance, in Nizamabad district, the decline in tank irrigated area is very sharp (Fig. 8.3); an inverse linear relationship exists between tank irrigated area and well irrigated area; and the regression value very high ($R^2 = 0.55$ for both total well irrigated area and bore well irrigated area). For every one hectare increase in well irrigated area, the tank irrigated area declined by 0.25 ha. The relationship was quite week in the case of Vijayanagaram district. Interestingly, no relationship seems to exist between net tank-irrigated area and net well irrigated area (and also net bore well irrigated area) in the case of Kurnool district. In fact, the net area under tank irrigation did not show any consistent decline in the district,

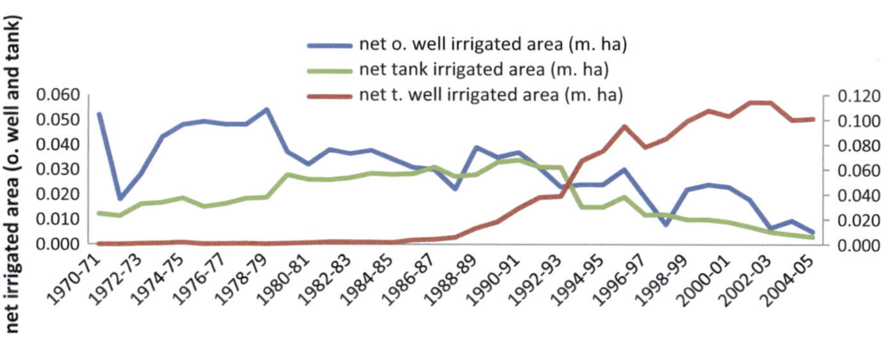

Fig. 8.3 Historical change in area irrigated by different sources, Nizamabad

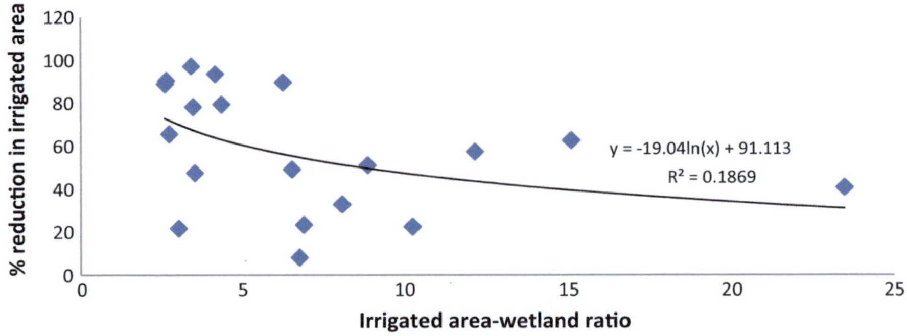

Fig. 8.4 % Reduction in tank irrigated area versus 'irrigation-wetland ratio'

which has large tracts of forest land and many tanks, which have forested catchments.

Further analysis was carried out to examine whether any relationship exists between tank characteristics, defined by the ratio of the area irrigated by the tank and its wetland area, and the degree of reduction in irrigated area over time, which these tanks have undergone. The analysis was carried out using district as the unit. The percentage reduction in irrigated area was estimated for each district by taking the ratio of the reduction in area during 1970–71 to 2004–05 and dividing it by the area irrigated in 1970–71. Here, we assume that the area irrigated by the tank in 1970–71 reflect the best irrigation performance for the tanks in all the districts. Our analysis shows an inverse (logarithmic) relationship between the "irrigation-wetland area ratio" and the reduction in area which irrigation from the tank has undergone, defined in percentage terms (Fig. 8.4). Greater the value of the wetland-irrigation ratio, lower the reduction in area irrigated over time. Frequency analysis showed that tanks with wetland-irrigation ratio in the range of 2.0–5.0 experienced an average reduction in irrigated area of 73.5%. Against this, those tanks having wetland-irrigation ratio exceeding 5.0 (between 6.26 and 23.47) experienced an average reduction in the irrigated area of only 48.70%. Many districts in this category showed less than 25% reduction in irrigated area. These analyses indicate that the tanks with low irrigation-wetland ratio are likely to deteriorate much faster than those with high irrigation-wetland ratio.

8.8 Micro Level Analysis of Impact of Catchment Land Use on Tank Performance

The impact of changes in land use in the catchment on the performance of tanks was analyzed by comparing temporal changes in total area under different crops and tank-irrigated area under different crops in different seasons in the command area against the temporal changes in types and density of groundwater abstraction structures and cropping in the catchment. The impact of land-use changes on tank performance was also analyzed by comparing the performance of two tanks in terms of percentage area under irrigation and irrigated area ratio at a given point of time against the characteristics of land use in the catchment of the tanks such as well density and area under cropping at that point of time.

The analysis was done for both tank level data and individual farmer level data. The tank level data included: (a) area under different crops in different seasons in the command area and catchment area; (b) area of different crops irrigated by tank water in different seasons in the command area; and, (c) density of wells in the command area and catchment area.

The summary of analysis for the tank-level data is provided in Table 8.2. It covers the tanks which are performing well and tanks which are not so well performing from all the three districts. Two types of comparison in tank performance were possible. (1) The difference in temporal performance of tanks located in different districts. (2) Difference in historic performance of tanks from the same location (district). This is compared against the physical and socio-economic features of the tanks under study to understand the reasons for differential performance.

Table 8.2 provides the data on performance of the six tanks in terms of gross cropped area and gross irrigated area over time. The main highlights are as follows.

Tanks in Kurnool are the best tanks in terms of the overall condition as it had sufficient storage even in the beginning of summer season. The catchment area of the tank is covered by reserve forests and as such has no cultivation or no wells. Though the irrigated area ratio (ratio of gross irrigated area against the command area) is small for the first tank (ranging from 0.53 in 1980 to 0.44 at present), a large amount of water from the tank is diverted for domestic water supplies. The survey showed that in addition to irrigation, the 300 families from the village also use tank water for livestock drinking. Second: the command area farmers of the first tank (relatively better condition) grow water intensive crops such as chilly along with paddy and jowar, whereas farmers in other district tanks growth low water consuming short duration pulses such as green gram and black gram, which largely uses residual soil moisture from the harvested paddy fields. Further, unlike in the case of both the tanks in Nizamabad and one of the tanks Vizianagaram, there are fewer wells in the command of the tanks surveyed in Kurnool.

As Table 8.3 shows, all the 26 farmers surveyed in the command were found to be using only tank water in Parumanchala tank. In the case of Padmaraja tank, only two of the 33 farmers surveyed were using well water. Hence, it can be very well assumed that in the case of Kurnool tanks, a remarkable share of the irrigation in the command is from tank water. In the case of Nizamabad, the well density in the command area of the degraded Jukkul tank has been increasing over time, and is very high now. Twenty two out of the 49 farmers surveyed from the tank command reported using well in conjunction with tank water while 12 of them use only well water. Hence, most of the irrigation reported in the command area must be from wells only. Therefore, though the irrigated area ratio for the tank command has been increasing, using that alone for comparing tank performance will be highly

Table 8.2 Tank Performance Vs Tank Characteristics

Name of district	Tank No.	Good/degraded	Name of the tank	Command area (ha)	Gross cropped area in the command (ha)					Gross irrigated area in the command (ha)				
					1970	1980	1990	2000	Current	1970	1980	1990	2000	Current
Kurnool (A)	1	Good	Parumanchala	607.0		324.0	301.7	313.9	269.3		324.0	301.7	313.9	269.3
	2	Degraded	Padmaraja Tank	344.3			368.6	405.0	328.1			368.6	405.0	328.1
Nizamabad (B)	1	Good	Gundla Tank	122.31	121.41	141.75	141.75	170.10	202.50	121.41	141.75	141.75	170.10	202.50
	2	Degraded	Jukkul Tank	63.99	64.80	70.47	70.47	70.88	95.18	64.80	70.47	70.47	70.88	95.18
Vizianagaram (C)	1	Good	Pedda Tank	303.51	526.50	567.00	567.00	526.50	546.75	413.10	405.00	425.25	425.25	441.45
	2	Degraded	Nalla Tank	127.48	255.15	255.15	255.15	255.15	145.80	255.15	255.15	255.15	255.15	137.70

Table 8.3 Percentage of sample farmers using different sources of irrigation in the tank commands

Name of source	Percentage of farmers using different sources of irrigation in the command of					
	Nizamabad		Kurnool		Vizianagaram	
	Gundla tank	Jukkul tank	Parumanchala tank	Padmaraja tank	Pedda tank	Nalla tank
Tank	69.0	24.5	100.0	84.8	71.4	100.0
Open well		12.2		6.1	3.6	
Tank and open well		16.3			7.1	
Bore well	11.9	6.1		3.0	10.7	
Tank and bore well	16.7	6.1			3.6	
Rain-fed		8.2		6.1	3.6	
Tank and rain	2.4	22.4				
Open well and bore well		2.0				
Open well and rain		2.0				

Source Authors' own analysis using primary data from tank water users

misleading. Interestingly, though Nalla tank is highly degraded, in the absence of wells, the farmers in the command only depend on tank water. Hence, the bad condition of the tank is reflected in the irrigated area, which has reduced over time.

Comparison between two tanks in the same location brings out the effect of catchment land use changes in a better way. In the case of Gundla tank in Nizamabad and Pedda tank in Vizianagaram, which were perceived as in good condition by the local people, the irrigated are ratio has increased over time. For instance, in the case of Gundla tank, it increased from 0.99 in 1970 to 1.66 at present, whereas in the case of Pedda tank, it increased from 1.36 in 1970 to 1.45 at present. This is quite contrary to what was found in the districts in terms of historical performance of tanks. In both the districts, the tank irrigated area declined, while the decline was very drastic in the case of Nizamabad.

What makes these tanks distinct is the fact that there is no significant groundwater use in their catchments. While there are no wells in the catchment area of Gundla tank (tank of Nizamabad, which is in relatively better condition), there are very few wells in the catchment of Pedda tank (1 well per 20 ha of catchment). But, this increase in irrigated area cannot be attributed to the good condition of tanks alone. The current groundwater use in the command area is significant for winter and summer crops, though these are low water consuming crops. As seen during the field work in Vizianagaram, almost all farmers in the command area of Pedda tank have bore-wells with electricity connections. Wells are the main source of water for irrigation of winter crops (maize) and summer crops. Even during the month of summer, a significant portion of the command area is under sesame, which is a short duration crop.

Contrary to these, in the case of Jukkul tank (degraded tank of Nizamabad), there are around 37 wells per 20 ha of the catchment area, as on today, which actually experienced an increase from around 13 wells per 20 ha in 1970 to 20 wells per 20 ha in 1990. Further, the area under cultivation in the catchment as on toady is 90%, and it increased dramatically from 38.45 ha to 182.1 ha over a period of 40 years.

Though there are no wells in the catchment of Nalla tank (degraded tank of Vizianagaram) in Vizianagaram district, rain-fed cultivation in the catchment has drastically increased over time—with the area increasing to 188 ha at present, covering 37% of the catchment area. Probably due to this, the irrigated area ratio for the tank declined drastically from 2.0 to 1.1 during the period from 2000 to 2011. This means that the actual performance of tanks in terms of area served by it is much lower than the reported (irrigated) area in the tank command and in reality, might have declined.

8.9 Agricultural Activities in the Command and Catchment of Tanks

In this section, we would analyze the changes in crop production in both the commands and catchments of the tanks. This is based on primary data collected from 25 farmers located in the command and catchments from each tank location, with a total of roughly 150 farmers from the six tanks selected for investigation. While the changes in cropping pattern and cropped area in the catchment is likely to influence the tank hydrology, the changes occurring in the cropping pattern and cropped area in the command area is a reflection of the changes in tank hydrology, and other farming related externalities which include the change in access to water from underground and other sources. As discussed in the first section, the scenario vis-à-vis access to groundwater had dramatically changed in Andhra Pradesh after the 80s with the advent of energized pump sets for extracting water from open wells, and bore wells that are able to tap water from deeper strata. If market conditions remain the same, reduced inflows into the tanks can force farmers to reduce the area under water intensive crops like paddy and sugarcane and shift to low water consuming crops like pulses. But, improved access well water, through energized pump sets and drilling technologies can enable farmers to intensify their cropping and go for irrigated crops during winter and summer seasons. The results of the analysis are presented district-wise and tank wise below.

8.9.1 *Kurnool District*

Tank 1: Parumanchala Tank, Parumanchala Village, Nadikotkur Mandal, Kurnool District

Table 8.4 shows the cropping pattern of Parumanchala tank command for the period from 1970 to 2011 (current). The analysis shows that there has been a remarkable increase in the cropped area in the command during the past 40 years, with most of the expansion occurring the first decade i.e., 1970–80. During this period, the cropping intensity also increased substantially, with crops being cultivated during winter and summer. Thereafter, it had hovered around 155–165 ha. Even during 2000–11, the change in gross cropped area was negligible. While the area under kharif paddy did not show any major change during 1970–2000, the percentage area under kharif paddy drastically declined (from 69% in 1970 to 4.8% in 2011), as new crops were introduced in the command. These new crops include chilly, groundnut, tobacco, summer jowar, cotton and pulses. Chilly occupies nearly 45% of the cropped area of the sample farmers, which is only second to jowar, which is the major kharif crop in the tank command today (Table 8.4).

The corresponding figures for the tank catchment are shown in Table 8.5. The data shows that cropping is practiced only during kharif season, and ground nut and red gram being the major crops. These crops require much less irrigation water as compared to paddy. Though these are also reported to be irrigated, the availability of rains means that they require very little irrigation that too during the dry spell.

Tank 2: Padmaraja Tank of Indireswaram Village, Atmakur Mandal

Table 8.6 shows the cropping pattern and irrigated cropping pattern in the command area of Padmaraja tank, which is considered to be a well performing tank in the area for the period from 1980 to 2011. Since all the crops are irrigated in almost all situations the cropped area figures and irrigated area figures are more or less the same, except in 1990. There has been no significant change in the cropped area in the tank command since 1980, except a 15% increase during 1990–2000 as compared to the base year of 1980. Thereafter, the area under cropping and irrigation reduced slightly. The tank command also did not witness much change in the cropping pattern also. The area under paddy hovered around 69–85% during the past 3 years. Unlike in the case of Parumanchala tank, where the percentage area under paddy reduced drastically after 1970, in this case, it went up slightly during 2000–11. Nevertheless, no major cropping was reported during the winter season, while at least four crops viz., cotton, sunflower, chilly, and jowar, were grown by farmers during the season during 1980–2000. As regards tank catchment, no crops were grown there, as it is under reserve forest.

Table 8.4 Percentage cropped and irrigated area under different crops of sample farmers in Parumanchala tank command

Season	Crop	Percentage area under different crops in				
		1970	1980	1990	2000	Last normal year
Kharif	Paddy	69.1	16.5	19.7	18.4	4.8
	Maize			3.3		4.5
	Jowar	11.4	11.8	15.0	17.9	34.4
	Cotton	19.4	1.1	1.3	1.2	1.2
	Chilli		25.3	23.0	38.7	27.0
	Groundnut		3.8	4.6	4.3	7.2
	Tobacco		8.2	9.8	9.1	9.0
	Sunflower		4.1	11.4	3.0	3.0
	Pigeon Pea		5.5	6.5	6.1	9.0
	Black Gram			3.9		
Rabi	Black Gram		1.1	1.3	1.2	
Summer	Jowar		0.8			
	Tobacco		8.2			
	Chilli		8.2			
	Pigeon Pea		5.5			

[a]*Note* The cropped and irrigated areas in Parumanchala tank command are one and the same and therefore the values under 'percentage area under different crops' in the table represent both

8.9.2 Nizamabad District

Tank 1: Jukkul Tank, Bhavanipet Village, Machareddy Mandal
A quick review of the data on gross cropped area and cropping pattern shows that there has been a significant reduction in the gross cropped area since 1970—with the gross cropped area declining from nearly 60 acres in 1970 to around 36 acres in the last normal rainfall year.

As regards the cropping pattern, no pattern seems to be emerging, except that during dry years, farmers reduce the area under both kharif and winter paddy. Another important observation is that sugarcane was introduced as a major crop during 1980s, but had almost disappeared by 2000. This could be attributed to the introduction of energized wells in the area, and increase in command area farmers' access to well water for irrigation. Currently, the farmers in the command do not grow sugarcane. Overall, reduction in cropped area is a strong indication of water shortage faced by tank irrigators.

Since Jukkul tank is a degrade tank, it is important to examine the changes in farming systems in the catchment area. The data shows that the area under cropping has increased in the command area, from 11.7 acre in 1970 to 21.20 acre in the last normal year (2011). An interesting observation is that sugarcane was introduced as a major crop as the data for 1990 indicate, but farmers did not continue this crop. Paddy continues to the dominant crop in the catchment, occupying around

Table 8.5 Gross cropped and irrigated area (acre) of sample farmers in Parumanchala tank catchment area

Name of season	Name of crop	1980		1990		2000		LNY	
		cultivated	irrigated	cultivated	irrigated	cultivated	irrigated	cultivated	irrigated
Kharif	Groundnut	24	24			24	24		
	Red gram			24	24			14	14
	Tobacco							10	10

Source Authors' own analysis based on primary data

Table 8.6 Percentage cropped and irrigated area under different crops of sample farmers in the Indireswaram tank command

Name of season	Name of crop	Percentage area of crops cultivated and irrigated in the Indireswaram tank command							
		1980		1990		2000		Last Normal Year	
		Cropped	Irrigated	Cropped	Irrigated	Cropped	Irrigated	Cropped	Irrigated
Khariff	Paddy	76.9	76.9	65.8	68.9	77.4	77.4	84.5	84.5
	Cotton	1.9	1.9	4.3	4.5				
	Sunflower	14.4	14.4	6.7	2.2	5.9	5.9	2.5	2.5
	Maize	1.3	1.3	1.4	1.5	1.1	1.1		
	Jowar			15.7	16.5				
	Jute					1.6	1.6	1.9	1.9
	Gooseberry							10.6	10.6
Rabi	Cotton	1.3	1.3	1.4	1.5	1.1	1.1		
	Sunflower	2.2	2.2	2.5	2.6				
	Chilly	0.6	0.6	0.7	0.7	11.4	11.4		
	Jowar	1.3	1.3	1.4	1.5	1.1	1.1		
	Watermelon					0.5	0.5	0.6	0.6

Source Authors' own analysis using primary data

80% of the total cropped area, though in 2000 maize was the major crop occupying around 75% of the gross cropped area.

As regards the irrigation in the command, there has been an overall reduction in the area irrigated by tank—from 57 acres in 1970 to 34 acres last year, though maximum shrinkage in irrigated cropped area was reported for 1990 (by nearly 2/3d). Thereafter, the irrigated area improved to become 45 acres in 2000 and then declined. In line with data on cropping pattern, maximum area under irrigated sugarcane was reported in 1990 (around 28 acres), occupying around 75% of the total irrigated area in the command. A useful observation vis-à-vis irrigated cropping pattern is of maize becoming a dominant crop during drought years. During the last drought, it was reported to have occupied around 86% of the total area under irrigation.

As regards the catchment land use, the irrigated area is more or less the same as that of cropped area, meaning all the crops cultivated in the catchment are also irrigated.

Tank 2: Gundla Tank, Domakonda Village and Mandal

Gundla tank is a better performing tank when compared to Jukkul tank. The outcomes of the analysis of data on cropped area and irrigated area of different crops in the tank command are presented in Table 8.7. In the case of Gundla tank, no notable and consistent reduction or increase in area reported by the farmers was seen over the years. What is more important is the fact that paddy remained as the most dominant crop in the command, raised during both kharif and winter season and occupying more than 80% of the cropped area.

8.9.3 Vizianagaram District

In the case of the two tanks selected from Vizianagaram district, no cultivation in the catchment was reported by farmers through the village survey showed some cultivation in the catchment area. As per the tank level data obtained from the village tank users' association, while no data on cropping are available for the present situation in the case of Pedda tank, no data on historical cropping (1970–2000) are available for Nalla tank. Though a lot of discrepancy is observed vis-vis data on land use in the catchment, the fact remains that there are no wells in the catchment area, which is suggestive of low intensity of land use there in terms of agriculture. Hence the outputs were generated only for cropped area and irrigated area in the tank command.

Tank 1: Pedda Tank, Rellivalasa

In the case of Pedda tank, which is considered to be one in good condition, the area under cropping had consistently increased from 1970 to 2000. But, later on the total cropped area of sample farmers declined to around 108 acres in 2011, the last normal year. The increase in cropped was also accompanied by increase in area under paddy, winter groundnut and summer maize. As regards the cropping pattern,

Table 8.7 Area under different crops (acre) and irrigation of sample farmers in Gundla tank command

Name of season	Name of crop	Area under the crop in					
		1970	1980	1990	2000	Last normal year	Last drought year
Kharif	Paddy	51.6	62.9	6.9	64.4	61.9	57.6
	Maize					5.0	9.0
Rabi	Paddy	48.2	62.9	57.0	41.5	45.9	26.0
	Maize			2.0	2.0	5.0	10.0
Total		99.8	125.9	66.0	107.9	117.8	102.6
Percentage area under different crops in							
Kharif	Paddy	51.7	50.0	10.5	59.7	52.6	56.1
	Maize					4.2	8.8
Rabi	Paddy	48.3	50.0	86.4	38.5	38.9	25.4
	Maize			3.0	1.9	4.2	9.8

Source Authors' own analysis

the percentage area under paddy, groundnut and maize did not fluctuate much over the years. The area under irrigation was also found to be same as that under cropping as all the crops were irrigated. Currently, farmers are found to be growing sesame during later summer, which lasts till the end of July.

Tank 2: Nalla Tank of Pinavemali, Vizianagaram Mandal

In the case of Nalla tank, the total cropped area of sample farmers surveyed in the command consistently decreased from 153.7 acres in 1970 to as low as 91 acres in 2000, and then increased to 165 acres in the last normal year (2011). But, a substantial portion of this increase in area came from pulses such as green gram and black gram, which are very low water consuming, short duration crops. The area under green gram and black gram went up by 52 acres. Though area under kharif paddy, which receives only supplementary irrigation, was very high during the last normal year (a rise of 38 acres from 2000 figures), there was reduction in area under winter maize, which is fully irrigated. The gross cropped area was, however, down to 116 acres in the last drought year. The percentage area under paddy did not change significantly during the forty year period from 1970 to 2011, though it was highest during 1990.

The irrigated area in the Nalla tank command is a little lower than the cropped area, with a small portion paddy is left un-irrigated. Paddy occupies nearly 45% of the cropped area. Nevertheless, during drought year, the percentage area under.

8.10 Performance of Tanks Against Their Physical and Socio-Economic Attributes

We started off with the basic premise that gross irrigation as a ratio of the command area (irrigated area ratio) is an important indicator of tank performance, and then went on to analyze the effect of groundwater irrigation and crop cultivation in the catchment on tank performance. But, our analysis shows that this can often be misleading due to two reasons.

First: the cropping pattern changes drastically from tank to tank, and often is a function of the tank hydrology itself. The farmers in the command area of tanks which receive sufficient inflows tend to grow highly water intensive crops such as chilly, sugarcane and sunflower, whereas those in the command area of tanks facing water shortage tend to grow low water consuming crops such as pulses after kharif paddy. From perennially water-rich tanks, water is also used for domestic water supplies in the neighbouring villages and small towns, apart from livestock drinking.

Second: the presence of wells in the tank command alters the scenario of irrigated area in the tank command. Even in instances where the tanks are not able to serve the command area farmers because of reduced inflows from the catchments, the wells in the command area meet the crop water requirement. But, under no circumstances, the presence of wells in the command influences the farmers not to use water from the tanks. Instead, it is the inability to get sufficient water from tanks for irrigating crops that motivates them to go for well irrigation. Therefore, high density of wells in the command area should be treated as a sign of low dependability of tank system as a source of water for irrigation and other uses in the command.

When the analysis looked at the variations in cropping pattern in the command area, the uses of tank water other than for irrigation (like domestic water supply to towns and villages and livestock drinking), and the effect of wells in the tank command on irrigation performance to do the comparative performance of tanks, the following becomes clear. The performance of those with high density of wells in the catchment area and high land use intensity are likely to decline drastically over time, whereas those without much interception of their catchments through faming and groundwater withdrawal sustain their performance, without any threat to the hydrological integrity. The best performing tanks will be those which have no cultivation in the catchment (which also implies that there are no wells), and the second best are those which have no wells in the catchment, but have rain-fed cultivation. As a result, tanks which have their upper catchments located in forests are the most ideal ones in terms of performance.

8.11 Findings

Several researchers have enquired into the causes for decline of tank irrigation in South India. They have largely attributed the "decline" of tanks to social and institutional factors. The most dominant of them are the lack of incentive among the command area farmers, and collapse of traditional management institutions, including community management structures and institutions of overlords (*Zamindars*) which took care of their upkeep, increase in groundwater irrigation and the consequent reduction of interest among farmers in tanks, interception of supply channels, and lack of adequate attention paid to regular maintenance.

The arguments about collapse of traditional institutions as the cause of "tank decline" are based on a resounding view that even the external factors, which have potential influence on the tank performance, were within the control of these institutions. On the contrary, there are several physical and socio-economic changes taking place in and around the tank systems over time, which could have impacted on tank performance, and they are not within the control of these institutions. They inadvertently ignore the fact that these institutions existed in a certain socio-political framework, which cannot be recreated. Again, such views are based on the assumption that simply cleaning supply channels, or clearing the catchments or repair of the tank embankments, and de-silting of the distribution network would yield results in terms of improved hydrological and ecological performance of and increased storage in tanks. So far as the well irrigation argument goes, it assumes that wells were within the reach of all segments of peasant, small and big. This again is not true as only a tiny fraction of the small and marginal farmers own wells, and still have great stakes in tanks.

These factors at best became contextual variables for tank deterioration, and not explanatory variable as the causality has not been tested. Conversely, it is probably the decline in tank performance due to extraneous reasons that had resulted in community's disinterest in their management, with the cost of maintaining them surpassing the actual benefits that can be accrued from their upkeep. We propose an alternative hypothesis that excessive groundwater draft characterized by groundwater irrigation in the tank catchment and commands, and land-use changes in the catchment in the form of intensive crop cultivation resulted in reduced tank inflows, causing decline in area irrigated by tanks. These hypotheses were tested using: (a) secondary data available from government agencies at the district level on area irrigated by tanks and wells; and (b) primary data collected from tank communities at the local level on changes in groundwater irrigation in catchments and commands, and changes in catchment land use.

Groundwater irrigation had grown steadily in Andhra Pradesh since the early 70s till the end of the last century. As our analysis shows, the net well irrigated area began to "plateau" after 2000–01. Thereafter, as pointed out by Kumar et al. (2011), the increase in number of wells had not resulted in increase in gross well irrigated area either.

Analysis at the level of the state and districts showed that increased groundwater draft adversely affected the performance of tanks, as indicated by the strong correlation between well and bore well irrigated area and net area irrigated by tanks at the state level for AP, and also for many districts of the states. Therefore, it could be safely argued that much of the expansion in well irrigated area happened at the cost of tank irrigation. Nevertheless, the effect of well irrigation on tank performance hasn't been uniform. While in many districts, the decline in tank performance in terms of "net area irrigated by tanks" in response to increase in "net well irrigated area" has been sharp, in some districts, it has been mild.

In lieu of the adverse impact of well irrigation on tank performance, it could be stated that though the net increase in well irrigated area in the state has been a remarkable 1.4 m. ha (net), the overall contribution of wells to expansion of irrigation in the state will be much less, if one takes into account the fact that the reduction in net tank irrigated area is around 0.60 m. ha, i.e., from 1.11 to 0.47 m. ha. But, the reduction in tank irrigated area cannot be fully attributed to groundwater over-extraction, and part of the reduction might have been caused by the change in land use in the tank catchments.

Analysis carried out to examine the relationship exists between tank characteristics, and the degree of reduction in irrigated area over time showed an inverse (logarithmic) relationship between the "irrigation-wetland area ratio" and percentage reduction in irrigated area of the tank, defined. Greater the value of the wetland-irrigation ratio, lower the reduction in area irrigated over time. Frequency analysis showed that tanks with wetland-irrigation ratio in the range of 2.0–5.0 experienced an average reduction in irrigated area of 73.5%. Against this, those tanks having wetland-irrigation ratio exceeding 5.0 (between 6.26 and 23.47) experienced an average reduction in the irrigated area of only 48.7%. Many districts in this category showed less than 25% reduction in irrigated area. These analyses suggest that the tanks with low irrigation-wetland ratio are likely to deteriorate much faster than those with high irrigation-wetland ratio.

As regards the impact of catchment land use on tanks, some of them are positive, some of them are negative. The first type of change is in the area under cultivation in the catchment. The catchments of tanks are generally public land with government forests, pasture land and revenue wasteland. Barring the reserve forests, these catchments are increasingly being encroached-upon by individual villagers for cultivation. One factor which triggered intensive land use in the catchment is the access to well irrigation. With water available from wells for supplementary irrigation, farmers are able to take up cultivation of kharif crops even in the driest regions of the country. Hence, intensive groundwater irrigation in the catchment had double impact on tank hydrology, firstly by affecting the groundwater outflows into streams and the secondly by affecting the runoff from the catchment entering the tank reservoir.

Catchment cultivation alters the hydrology by reducing the runoff generation potential of the incident rainfall and the impoundment of part of the generated runoff into the cultivated fields through the field and farm bunds. Often, afforestation activities are undertaken in the catchment by community organizations, which affect runoff generation. The trees such as eucalyptus which were preferred for plantation under afforestation programme were water-guzzlers. They can guzzle groundwater apart from capturing part of the runoff and depleting soil moisture. Such changes are occurring everywhere in the rural landscape. But, there are two notable exceptions. (1) Places where the forests constitute the main catchment of tanks. (2) Areas where groundwater development is not feasible due to presence of hard rock formation. Thus, in areas which have experienced significant changes in land use in the form of expansion of cultivation in the catchment, it won't be economically prudent to invest in tank rehabilitation.

The second type of change in the catchment land use is caused by the use of clayey soils in the catchment and tank bed for brick making etc. With booming construction activity in the state, there is mounting demand for bricks. The pressure on catchment land and tank bed for such uses is more in case of tanks which are located in the vicinity of towns. Such activities can also change the runoff or storage potential of the tank, depending on the place from where the soil is excavated. In such situations, the communities or the local Panchayats will not have much interest in reviving the tanks as the income earned from such activities is very large.

The third type of change is the interception of the drainage lines in the catchment. There are two different types of activities which cause this interception. (1) Construction of roads and buildings. (2) The indiscriminate construction of water harvesting structures such as check dams. This is very rampant in Andhra Pradesh, like in many semi-arid states of the country. The absence of any kind of regulations on water resources in the state had actually precipitated in serious concerns. One of the reasons for this unprecedented increase in water harvesting schemes is that the demand for water in agriculture has increased in the upper catchments of river basins with growing population pressure, and with depleting groundwater resources. In such areas, where such inflow reduction is clearly visible, it won't be economically prudent to invest in large-scale tank rehabilitation.

The degradation of tanks occurring as a result of changes in tank hydrology also seems to affect the success of the rehabilitation programme. It was observed that in the case of Kurnool and Nizamabad districts, more money was spent for rehabilitation works of those tanks which are actually not performing well as compared to the good ones. In spite of this, the condition of poorly performing tanks did not improve.

In the final analysis, it appears that groundwater intensive use in upper catchment or lower catchment will have the most remarkable impact on hydrology and performance of tanks. This is because, pumping of water from wells while reducing the groundwater outflows into streams, also leads to intensive use of land in the catchment for cultivation, which further leads to reduction in runoff generation and in situ harvesting of a portion of that runoff for storage in soil profile. This has affected hydrological and ecological integrity of many tanks, especially in water scarce basins. In nutshell, tank management is increasingly becoming a hydro-institutional challenge.

8.12 Protocol for Tank Rehabilitation

- Currently, no clear protocol exists for choosing tanks for rehabilitation. Prior to the selection of any tank for rehabilitation, the potential of its catchment to yield sufficient water as inflows into the reservoir needs to be ascertained. The estimate of catchment yield should be compared against the total water demand for competitive uses that exist for the tank under consideration. Only those tanks which receive sufficient inflows to serve the full command area at least once in three years, need to be taken up for rehabilitation. Tanks which receive inflows to the full storage capacity or sufficient water to meet the demand of the design command area in three out of four years, should receive highest priority for rehabilitation.
- But since this vital hydrological data on stream flows are not available for most tanks, it is imperative to study the land use in the tank catchment and command thoroughly, before embarking on rehabilitation of the tank system. It is quite evident that for catchments which have undergone major changes in land use, new models for runoff estimation will have to be developed, instead of using old rainfall-runoff models developed on the basis of observed stream flows of the past, when land use and land cover were drastically different.
- The most pragmatic approach for estimating dependable yield from tank catchment would be to estimate runoff for historical rainfall for the current land use in the catchment using US Soil Conservation Bureau's Curve Number method, and then estimate the runoff of certain dependability. In many situations, it may be necessary to redesign the tank command on account of the changing land use in the catchment and tank hydrology, and the new cropping systems which farmers have introduced in the area.
- For estimation of water demand for irrigation, kharif paddy in the entire command, and a short duration winter crop in 1/4th of the command assumed reasonably. For this, the existing cropping pattern of the area and the most desirable cropping pattern from the point of view of agro climate should be used as the basis. Since the irrigation water demand for paddy would vary widely between a drought year and a wet year, the amount of irrigation required during a normal year could be considered for demand estimation.
- In the case of cascade tanks, the upper catchment tanks should receive highest priority, if there are in need of repair or rehabilitation. This is because they are least likely to face problems of encroachment, change in land use in the catchment and intensive groundwater development in the catchments and command areas, by being located in forest catchments.

8.12.1 Quantitative Criteria for Choosing Tanks for Rehabilitation

1. Micro level studies to ascertain the feasibility of rehabilitation of tanks involves significant costs. Therefore, it is necessary to evolve simple and quantitative criteria for short-listing tanks for conducting detailed investigation, to finally decide on the nature of rehabilitation.
2. The analysis suggests that tanks which have relatively higher "irrigation-wetland area ratio should be given priority as they are likely to experience lesser degree of deterioration. One of the hydrological explanations for this is that such tanks would have higher losses from percolation as a result of increased groundwater draft. In areas where fishery is a major economic activity preferred by the tank communities, then the minimum water required in the tank for fish production should be considered instead of water requirement for winter crop production, for assessing the suitability for rehabilitation work.
3. Tanks with low density of wells in the command area and catchment area and low intensity of land use in the catchment need to be given priority while choosing tanks for rehabilitation. While it is difficult to arrive at a quantitative criterion for selecting tanks for rehabilitation based on number and density of wells in the catchment, a well density of one well per ha of command is a clear sign of diminishing importance of tanks for the farmers' livelihoods. On the other hand, a well density of 0.5 and above, i.e., one per two ha of land, is a clear indication of high intensity of land use in the catchment area, with multiple cropping. This can reduce the inflows into the tanks drastically by capturing the runoff in situ, and reducing the groundwater outflows into surface streams.
4. Large number of wells in the tank command may also indicate declining importance of tanks in the irrigated livelihoods of farmers in the command. The argument that farmers abandon tanks because of wells is also not true. Nevertheless there can't be a standard norm on well density and catchment land use, for selecting or rejecting tanks for rehabilitation. The effect of well density and catchment land use intensity on tank inflows would also depend on the rainfall of the region and the physiographic features of the catchment. In high rainfall regions with steep catchment slopes, the norm could be relatively flexible, whereas in semi-arid and arid regions with low to medium rainfall, the norm will have to be more stringent.

8.13 Conclusions and Recommendations

South Indian tanks form important wetland ecosystems. Groundwater intensive use in the catchment affects tank ecology by reducing the base flows that constitute a major part of the tanks inflows, and also reduces the physical performance of tanks in terms of the area irrigated. Though there are factors other than area irrigated, which need to be considered for assessing the performance of tanks, for a temporal study this could form a useful indicator of tank performance. But, the adverse impact of intensive well irrigation on tank performance in terms of reduction in tank irrigated area hasn't been uniform across districts.

Intensity of land use and density of wells in the catchment are major determinants of tank performance. Low density of wells and low land use intensity in the catchment provide favourable conditions for sustaining the hydrological and ecological integrity of tanks and therefore their good performance. Increasing intensity of wells and expansion in area under cultivation in the catchment reduces the irrigation performance of tanks, by reducing the surface runoff and base flows. Reduction in groundwater outflows into streams also impacts on the ecological health of the tanks, as the lean season flows into the tanks are mainly from base flows.

Managing tanks is increasingly becoming a hydro institutional challenge. The approach of taking up every tank for rehabilitation, on the justification that a local institution could be created to manage it, needs a critical review. This is because there are many external factors responsible for deterioration of tanks that are beyond the capacity of these institutions to control. The protocol for tank rehabilitation should include estimation of the actual yield potential of the tank catchment using rainfall-runoff models. This should be followed by realistic estimation of the water demand in the command, taking into account the existing cropping pattern and the variations in crop water demands in lieu of the inter-annual variability in rainfall. In the case of cascade tanks, the upper catchment tanks should receive highest priority in rehabilitation programmes. Since such studies for examining the feasibility of tank rehabilitation involve significant costs, some simple and quantitative criteria were identified to shortlist tanks for detailed studies which would help in investment decision making on rehabilitation.

References

Asian Development Bank (2006) Rehabilitation and management of tanks in India: a study of select states. ADB, Philippines

Balasubramanian R, Selvaraj KN (2003) Poverty, private property and common pool resource management: the case of irrigation tanks in south India. Working paper no. 2. South Asian Network for Development and Environmental Economics, Kathmandu, Nepal

Bassi N, Kumar MD, Sharma A, Pardha-Saradhi P (2014) Status of wetlands in India: a review of extent, ecosystem benefits, threats and management strategies. J Hydrol: Reg Stud 2:1–19

Dhawan BD (1985) Output impact according to main irrigation sources: empirical evidence from four selected states. Paper presented at INSA National Seminar on Water Management—The Key to Development of Agriculture, 20–30 Apr 1986

Easter KW, Palanisami K (1986) Tank irrigation in India and Thailand: an example of common property resource management. Staff paper P86-35. Department of Agricultural and Applied Economics, University of Minnesota, Minnesota

Gunnell Y, Krishnamurthy A (2003) Past and present status of runoff harvesting systems in dryland peninsular India: a critical review. Ambio 32:320–324
Juliano K, Simonovic SP (1999) The impact of wetlands on flood control in the red river valley. Natural Resource Institute, University of Manitoba, Manitoba, Canada
Kajisa K, Palanisami K, Sakurai T (2004) Declines in the collective management of tank irrigation and their impact of income distribution and poverty in Tamil Nadu, India. Foundation for Advanced Studies on International Development, Japan
Karthikeyan C (2010) Competition and conflicts among multiple users of tank irrigation systems. In: Fourteenth international water technology conference. International Water Technology Association, Alexandria, Egypt, 21–23 March 2010, pp 837–852
Kumar MD (2002) Making the montage: setting the agenda and priorities for water policy research in India, Report of the Annual Partners' Meet of IWMI-Tata Water Policy Research program, 19–20 Feb 2002
Kumar MD (2007) Groundwater management in India: physical, institutional and policy alternatives. Sage Publications, New Delhi
Kumar MD, Ghosh S, Patel A, Singh OP, Ravindranath R (2006) Rainwater harvesting in India: some critical issues for basin planning and research. Land Use Water Resour Res 6(1):1–17
Kumar MD, Patel A, Ravindranath R, Singh OP (2008) Chasing a mirage: water harvesting and artificial recharge in naturally water-scarce regions. Econ Polit Weekly 43(35):61–71
Kumar MD, Sivamohan MVK, Niranjan V, Bassi N (2011) Groundwater management in Andhra Pradesh: time to address real issues, Occasional Paper # 4. Institute for Resource Analysis and Policy, Hyderabad
Kumar MD, Panda R, Niranjan V, Bassi N (2013) Technology choices and institutions for improving economic and livelihood benefits from multiple uses tanks in western Orissa. In: Kumar MD, Sivamohan MVK, Bassi N (eds) Chapter 8: Water management, food security and sustainable agriculture in developing economies. Routledge, Oxford, UK
Mosse D (1999) Colonial and contemporary ideologies of 'community management': the case of tank irrigation development in south India. Modern Asian Stud 33:303–338
Muralidharan D, Athavale RN (1998) Artificial recharge in India, base paper prepared for Rajiv Gandhi National drinking water mission, Ministry of Rural Areas and Development, National Geophysical Research Institute
Narayanamoorthy A (2007) Tank irrigation in India: a time series analysis. Water Policy 9:193–216
Palanisami K (2006) Sustainable management of tank irrigation systems in India. J Dev Sustain Agric 1:34–40
Palanisami K, Easter KW (1991) Hydro-economic interaction between tank storage and groundwater recharge. Indian J Agr Econ 46(2):174–179
Palanisami K, Meinzen-Dick R, Giordano M (2010) Climate change and water supplies: options for sustaining tank irrigation potential in India. Econ Polit Wkly 45:183–190
Paranjape S, Joy KJ, Manasi S, Latha N (2008) IWRM and traditional systems: tanks in the Tungabhadra system. STRIVER policy brief no. 4. NIVA/Bioforsk, Norway
PRADAN (1996) Resource management of minor irrigation tanks and Panchyati Raj. Paper presented at the Seminar on Conservation and Development of Tank Irrigation for Livelihood Promotion, Madurai. 12 Jul 1996
Prinz D (2002) The role of water harvesting in alleviating water scarcity in arid areas. In: Key Note Lecture, Proceedings of the international conference on water resources management in arid regions, vol III. Kuwait Institute of Science Research, 23–27 Mar 2002, pp 107–122
Rao GB (1998) Harvesting water: irrigation tanks in Anantapur. Wasteland News 13(3)
Sakthivadivel R, Gomathinayagam P, Tushaar S (2004) Rejuvenating irrigation tanks through local institutions. Econ Polit Wkly 39:3521–3526
Sakurai T, Palanisami K (2001) Tank irrigation management as a local common property: the case of Tamil Nadu, India. Agric Econ 25:273–283
Sekar I, Palanisami K (2000) Modernised rainfed tanks in South India. Productivity 41(3):444–448

Shah Esha (2008) Telling otherwise a historical anthropology of tank irrigation technology in South India. Technol Culture 49(3):652–674

Shankari U (1991) Tanks: major problems in minor irrigation. Econ Polit Wkly 26:A115–A124

Sharma A (2003) Rethinking tanks: opportunities for revitalizing irrigation tanks-empirical evidence from Anantapur District, Andhra Pradesh, India, Working Paper # 62, International Water Management Institute, Colombo, Sri Lanka

von Oppen M, Rao KS (1987) Tank irrigation in semi-arid tropical India: economic evaluation and alternatives for improvement. Research Bulletin no. 10. International Crops Research Institute for the Semi-Arid Tropics, Patancheru, Andhra Pradesh

Part III
Moving Forward

Chapter 9
Interlinking Ecosystems for Multiple Benefits

Harpinder Sandhu

Humans depend on natural and managed ecosystems for the supply of ecosystem goods and services (ES) as raw material for industry and for human welfare (MEA 2005; Naeem et al. 2009). However, increasing population, changing land use patterns, growing food demand, has affected ecosystems globally, thereby impacting their services. The Millennium Ecosystem Assessment (MEA) sponsored by the United Nations assessed the consequences of ecosystem change for human well-being and provided a framework to identify and classify ES (MEA 2005). According to MEA, ecosystems sustain human life through the provision of four types of ES, supporting (e.g., pollination, nutrient cycling), provisioning (e.g., food production, fuelwood), regulating (e.g., water purification, erosion control), and cultural (e.g., aesthetic values). These ES not only perform life supporting functions necessary for human survival but also form the basis of most of the economic activities ranging from agricultural production, power generation, and other natural resource based industry. This concept of ES is widely accepted and has been used in different studies (Costanza et al. 1997; Daily 1997; Sandhu et al. 2008). ES classification developed by MEA has been applied for valuation of natural capital, accounting purposes, spatial analysis and also to develop mechanisms to pay for the protection of ES (Fisher et al. 2009; Kumar and Thiaw 2013). Building on the ecosystems theory and literature since the MEA, this chapter focuses on South Asian geographic region by exploring the importance of linking ecosystems for multiple benefits to sustain human well-being.

The importance of ES in supporting human life and as a life-support system of the planet (Daily 1997; Costanza et al. 1997, 2014; MEA 2005; Wratten et al. 2013) is now very well established and ES have been demonstrated to be of very high economic value (US$125 trillion/year; Costanza et al. 2014). Despite their importance, most of these services are not traded in economic markets and they carry no

H. Sandhu (✉)
School of the Environment, Flinders University, Adelaide, Australia
e-mail: Harpinder.Sandhu@flinders.edu.au

H. Sandhu (ed.), *Ecosystem Functions and Management*,
DOI 10.1007/978-3-319-53967-6_9

'price tags' (no exchange value in spite of their high use value). Therefore, there is no market signal that could alert society to changes in their supply. This has led to a trend of decline in ES globally with 60% of the ES examined having been degraded in the last 50 years (MEA 2005). Global efforts to halt this decline in ES have increased considerably since the completion of the Millennium Ecosystem Assessment (MEA) in 2005. The United Nations has established the Intergovernmental Science-Policy Platform on Biodiversity and Ecosystem Services (IPBES) to translate science into action worldwide in consultation with governments and research partners (IPBES 2010).

It is well known that ecosystems contribute substantially to the welfare of human societies by providing highly demanded and valuable ES (Kumar and Reddy 2007). Many of these ES, however, remain outside the conventional markets (Costanza et al. 2014). This is especially the case of public goods (non-timber forest products, food resources etc.) and externalities related to the active protection and management of resources. To capture this aspect, this chapter elaborates the ES approach that integrates social, economic and environmental context for sustainable management of ecosystems and their resources. It also provides valuation methods for assessing ES. It then develops a conceptual framework to integrate different ecosystems discussed earlier in the book for the sustainable well-being of the society.

9.1 Ecosystem Services Approach

An ES approach is one that integrates the ecological, social and economic dimensions of natural resource management (DEWHA 2009).

- It identifies and classifies the benefits that people derive from ecosystems, including market and non-market, use and non-use, tangible and non-tangible benefits.
- It describes and communicates these benefits in concepts and language that a wide range of stakeholders and the public can understand;
- It explores linkages between sustainable management of ecosystems and human welfare, including

ES approach includes description and categorisation of all ES, ecosystem based adaptation and economic values associated with them. These are discussed below.

9.1.1 Ecosystem Services Description

To capture the diversity of ES, MEA groups them into four broad categories based on their functional characteristics (see definition in Chap. 1). The four categories are defined below.

1. Provisioning Services: the provisioning function of ecosystems supplies a large variety of ecosystem goods and other services for human consumption, ranging from food in agricultural systems, raw materials and energy resources.
2. Regulating Services: ecosystems regulate essential ecological processes and life support systems through bio-geochemical cycles and other biospheric processes. These include climate regulation, disturbance moderation and waste treatment.
3. Cultural Services: ecosystems provide an essential 'reference function' and contribute to the maintenance of human health and well-being by providing spiritual fulfilment, historical integrity, recreation and aesthetics.
4. Supporting Services: ecosystems also provide a range of services that are necessary for the production of the other three service categories. These include nutrient cycling, soil formation and soil retention.

9.1.2 Ecosystem Based Adaptation (EbA)

An ES approach includes ecosystem based adaptation (EbA) that integrates biodiversity and ES into overall adaptation strategy to help people to adapt to the adverse effects of climate change (Colls et al. 2009). EbA is critical to implement sustainable ecosystem management practices and enhance resilience to climate change in developing countries. It is management tool that can be applied at different geographical scales (local, regional, national) and within various time frames (short to long term). It can be implemented as projects and as part of overall adaptation programmes. It is most effective when implemented as part of a broad portfolio of adaptation and development interventions (Colls et al. 2009). EbA can provide low cost and more accessible alternatives to rural or poor communities instead of measures based on infrastructure and engineering (Kumar and Thiaw 2013). It can integrate and maintain traditional and local knowledge and cultural values (Kumar and Kumar 2008).

EbA has been implemented at project level in both developing and developed countries at national, regional and local scales (Colls et al. 2009). EbA complements other responses to climate change by providing a full range of goods and ES, and provide proven and cost-effective protection against some of the threats that result from climate change (IPCC 2007). Some examples of EbA activities (CBD 2009; Colls et al. 2009) are,

- Coastal defence through the maintenance and/or restoration of mangroves and other coastal wetlands to reduce coastal flooding and coastal erosion.
- Sustainable management of upland wetlands and floodplains for maintenance of water flow and quality.
- Conservation and restoration of forests to stabilize land slopes and regulate water flows.
- Establishment of diverse agroforestry systems to cope with increased risk from changed climatic conditions.

- Conservation of agrobiodiversity to provide specific gene pools for crop and livestock adaptation to climate change.

9.1.3 Total Economic Value of Ecosystem Services

Total economic value distinguishes between use values and non-use values (Fig. 9.1; Barbier 2007; Barbier et al. 2009). Use values involve direct human use (consumptive or non-consumptive) whereas non-use values do not. Non-use values rely merely on the continued existence of an environmental resource and are unrelated to use. Use values can be either direct or indirect. Direct use values are more likely to have market value. Indirect use values, on the other hand, are difficult to quantify and are generally ignored in management decisions. For example, production of food is a direct use value for which money is exchanged in markets, whereas the nutrient cycling performed by soil organisms indirectly supports agricultural production but this contribution is non-marketed and goes financially unrewarded. Option values are value that are not currently used but have potential to be used in the future.

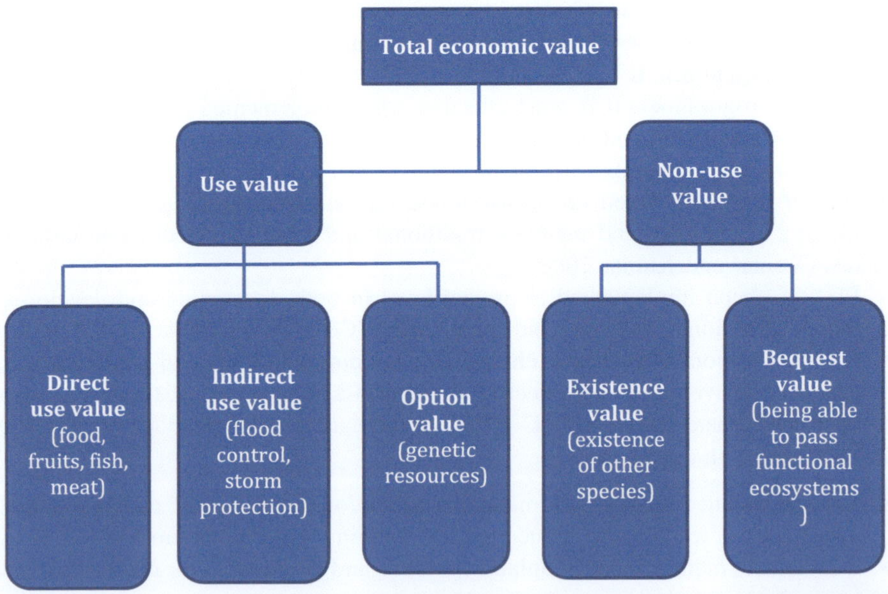

Fig. 9.1 The concept of total economic value (Pearce and Turner 1990; Barbier 2007; Barbier et al. 2009; De Groot et al. 2002; Balmford et al. 2008)

There are two ways in which people may value natural resources without using them either directly or indirectly. Existence value is the 'intrinsic' value of natural resources that arises simply due to their existence. It is the value that leads people to want to see other species like giant pandas, for example, protected, even if these people never see the species. Bequest value is what leads people to support policies and actions that protect natural resources for future generations to use.

9.2 Valuation of Ecosystem Services

Economic valuation methods fall into four basic types (De Groot et al. 2002; Kumar 2007, 2010), each with its own repertoire of associated measurement issues:

Direct market valuation: This is the exchange value that ES have in trade, mainly applicable to the 'goods' (i.e., production functions). Value of food produced is an example.

Indirect market valuation: When there are no explicit markets for services, we must resort to more indirect means of assessing values are needed. A variety of valuation techniques can be used to establish the (revealed) Willingness-to-Pay (WTP) or Willingness-to-Accept compensation (WTA) for the availability or loss of these services.

- Avoided Cost (AC): services allow society to avoid costs that would have been incurred in the absence of those services. Examples are flood control (which avoids property damages), waste treatment (which avoids health costs) by wetlands and biological control (which can lead to lower pesticide costs) of insect pests in vineyards.
- Replacement Cost (RC): services could be replaced with human-made systems; an example is natural waste treatment by marshes, which can be (partly) replaced with costly artificial water treatment systems.
- Factor Income (FI): many ES enhance incomes. An example is improvement of grapes quality which increase incomes of grape growers.
- Travel Cost (TC): Use of ES may require travel. The travel costs can be seen as a reflection of the implied value of the service. An example is recreation areas in vineyards and cellar sales that attract distant visitors whose value placed on that area must be at least what they were willing to pay to travel to it.
- Hedonic Pricing (HP): Service demand may be reflected in the prices people will pay for associated goods; an example is that housing prices in valley near vineyards usually exceed prices of identical inland homes near less attractive scenery.

Contingent valuation (CV): Service demand may be elicited by posing hypothetical scenarios that involve the description of alternatives in a social survey questionnaire. For example, such a questionnaire might ask respondents to express to their willingness to pay (i.e. their stated preference as opposed to revealed preference) improve vineyard restaurants so that they might enjoy outdoor sceneries.

Group valuation: Another approach to ecosystem service valuation that has gained increasing attention recently involves group valuation. Derived from social and political theory, this valuation approach is based on the principles of deliberative democracy and the assumption that public decision making should result, not from the aggregation of separately measured individual preferences, but from open public debate.

9.3 Society and Ecosystems

Human beings are an integral part of ecosystems and ES concept encapsulates the concept of three-dimensional sustainability—social, economic and environmental (Sandhu 2010). People living in rural or urban areas receive many benefits from land and water resources. They are dependent on different ES to fulfill basic needs and other constituents of well-being (Fig. 9.2; Sandhu and Sandhu 2014). These linkages can provide opportunities for the policy makers and development organisations to intervene by targeting each individual element of well-being and also enhance ES (Kumar 2007). There is potential to improve several well-being indicators by enhancing ES. For example, to fulfill food and nutritional demand, rural households need to produce adequate quantity of good quality food on their limited farmland. ES approach discussed above can help this by improving farm incomes using ecological techniques. Ecological techniques do not require a major farming system change but are based on enhancing ES in agriculture (Sandhu et al. 2015). For example, the role of leguminous crops in nitrogen fixation can reduce fossil

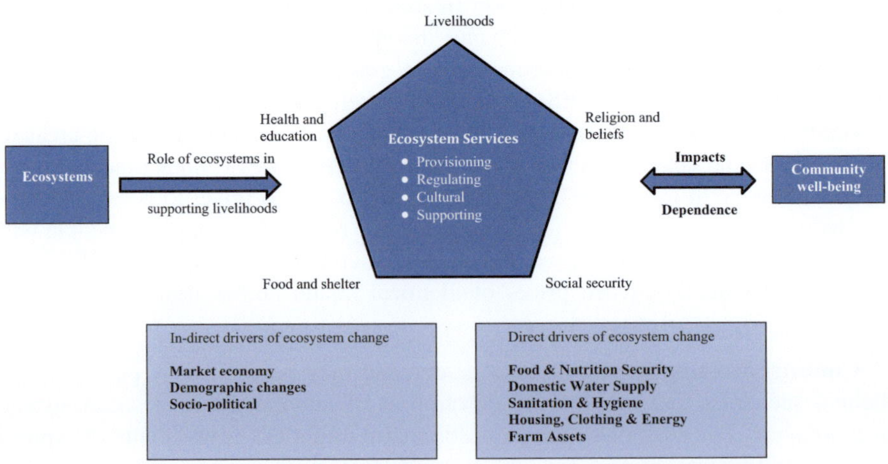

Fig. 9.2 Framework of ecosystems and well-being with drivers of ecosystem change

fuel-based fertilizer inputs (Gurr et al. 2004; Wratten et al. 2013). Similarly, sharing of common resources such as water by upstream and downstream users, fuelwood extraction and waste management can also be addressed by the local community at village level (Fisher et al. 2010).

There are often drivers of ecosystem change that may impact the ability of ecosystem to provide essential ES. These could be direct drivers that impacts resources in ecosystems. For example, growing demand for food can alter the land use in the forest ecosystem. Indirect drivers impact ecosystems from distant locations such as markets. For example, pursuit of supplying high value food products from farm has potential to accelerate land cover change thereby deteriorating local ecosystems (Geist and Lambin 2004).

9.3.1 Linkages Between Different Ecosystems for Human Well-Being

Human well-being comprising of social, economic and environmental aspects is dependent on healthy and functional ecosystems (Maslow 1943; Sen 2000; Sandhu and Sandhu 2014). Well-being is associated with the quality of life of an individual or group (Kumar and Kumar 2008; Summer et al. 2012). This could be related to fulfilling basic needs such as food, shelter, water etc. or other cultural and spiritual needs. Ecosystems, which are defined as a complex community of living and non-living components of the environment, provide many benefits for human well-being. The benefits derived from ecosystems in the form of goods or services are then known as ES. One condition of obtaining desired well-being is healthy and functional ecosystems. There is enormous complexity in ecosystems that support various life supporting functions and processes (De Groot et al. 2002). However, for simplification to be used in social and economic dimensions of human well-being, we divide them into different categories based on their specific function, ranging from marine based such as oceans, coastal etc. to land based such as forests, urban, rural, freshwater etc. (Kumar 2010). This chapter considers lands based ecosystems and which are defined in Chap. 1 (Mountain, agroecosystems, desert, forest, urban and fresh water). It provides a conceptual framework that explains how interlinkages between these ecosystems are necessary to sustain life and achieve human well-being.

9.4 Framework for Interlinking Ecosystems

All these ecosystems do not work independently of each other to support human life. Their functions depend on each other and they are inherently linked (De Groot et al. 2002; Bennett et al. 2009). As human life depends on many ES provided by these ecosystems, therefore there is need to develop an integrated approach to optimise the benefits out of these ecosystems (Turner and Daily 2008; Haines-Young and Potschin 2010). South Asia is a large part of earth's surface hosting a growing population with numerous demands on ecosystem and their services (Kumar and Reddy 2007). There is likely impact due to overuse of some of the resources. An integrated approach to linking ecosystems would thereby benefit both users and managers of these ecosystems for the long-term sustainability (Kumar 2010).

Human society is part of global ecosystems which comprises of terrestrial and marine ecosystems (MEA 2005). Terrestrial ecosystems can be divided further into agroecosystems, forests, mountains, desert and freshwater ecosystems. Human beings, whether living in urban areas or rural areas, are dependent on these ecosystems for their survival. Therefore, social, natural and built capital together provides human well-being through various ES (Fig. 9.3). Global economy operates within the global ecosystems and all economic activity is dependent on ecosystems for raw material or for other services (Washington 2012). Combination of different ecosystems results in providing benefits to human population which leads to human welfare through the provision of ES. Individual ecosystems can also provide limited benefits based on their functions. Therefore, to sustain future use of ecosystems and their services, there is need to consider integrated approach for the

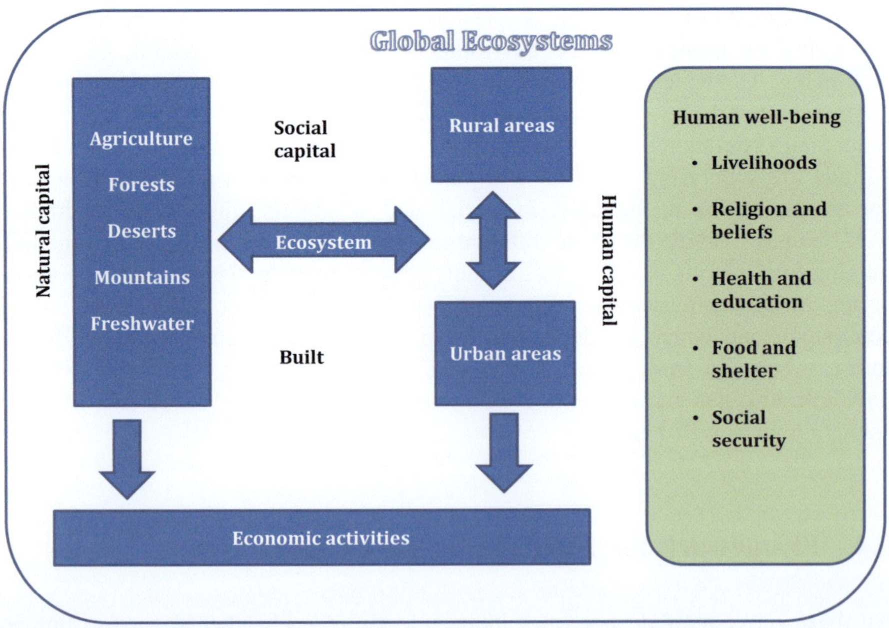

Fig. 9.3 Interlinking ecosystems for human well-being by providing linkages between natural, social, human and built capital

maintenance of these ecosystems. For example, there is increased influx of rural population to urban areas resulting in high demand of food and clean water in cities. Food can only be grown in agricultural areas; therefore, to maintain human population in cities, rural areas need to have a clear policy for sustainable development with matching investment so that they can continue to supply fresh and healthy food to cities. Similarly, waste recycling is major issue in cities and growing population puts more pressure on natural systems in this regard. This issue can be dealt with by maintaining the function of natural drains by ecological engineering. Similarly, forest, deserts and mountains provide essential ES for urban and rural population, supporting livelihoods and cultural aspects of well-being.

9.5 Summary

Societal dependence on ecosystems and their services is invaluable. Despite high social and economic value of these ecosystems, ES continue to decline bio-physically and monetarily. Recent estimates suggest that over last 15 years, there has been loss of ES in the range of US$4–20 trillion annually (Costanza et al. 2014). There is need to understand the functioning of ecosystem sand their importance to society so that policy can adopt new mechanisms to internalise the value of ES (Kumar 2007; Kumar and Thiaw 2013). This has potential to stop or slow down losses in the ES and degradation of ecosystems. Integrated approach as discussed above has potential to provide an alternative for the desirable and sustainable future of mankind.

References

Balmford A, Rodrigues A, Walpole M, Brink P, Kettunen M, Braat L, de Groot R (2008) The economics of ecosystems and biodiversity: scoping the science. European Commission, Cambridge

Barbier EB (2007) Valuing ecosystem services as productive inputs. Econ Policy 22:177–229

Barbier EB, Baumgärtner S, Chopra K, Costello C, Duraiappah A, Hassan R, Kinzig A, Lehman M, Pascual U, Polasky S, Perrings C (2009) The valuation of ecosystem services. In: Naeem S, Bunker D, Hector A, Loreau M, Perrings C (eds) Biodiversity, ecosystem functioning, and human wellbeing: an ecological and economic perspective. Oxford University Press, Oxford, pp 248–262

Bennett EM, Peterson GD, Gordon LJ (2009) Understanding relationships among multiple ecosystem services. Ecol Lett 12:1394–1404

CBD (2009) Connecting biodiversity and climate change mitigation and adaptation: Report of the second ad hoc technical expert group on biodiversity and climate change. Secretariat of the Convention on Biological Diversity, Canada

Colls A, Ash N, Ikkala N (2009) Ecosystem-based adaptation: a natural response to climate change. IUCN, Gland, Switzerland, p 16

Costanza R, d'Arge R, de Groot R, Farber S, Grasso M, Hannon B, Limburg K, Naeem S, O'Neill RV, Paruelo J, Raskin RG, Sutton P, vans den Belt M (1997) The value of the world's ecosystem services and natural capital. Nature 387:253–260

Costanza R, de Groot R, Sutton P, van der Ploeg S, Anderson SJ, Kubiszewski I, Farber S, Turner RK (2014) Changes in the global value of ecosystem services. Glob Environ Change 26:152–158

Daily GC (ed) (1997) Nature's services: societal dependence on natural ecosystems. Island Press, Washington, D.C.

De Groot RS, Wilson M, Boumans R (2002) A typology for the description, classification and valuation of ecosystem functions goods services. Ecol Econ 41:393–408

DEWHA (2009) Ecosystem services: key concepts and applications. Occasional Paper No 1. Department of the Environment, Water, Heritage and the Arts, Canberra, Canberra

Fisher B, Turner RK, Morling P (2009) Defining and classifying ecosystem services for decision-making. Ecol Econ 68:643–653

Fisher B, Burgess ND, Kulindwa K, Mwanyoka I, Turner RK (2010) Common pool resource management and PES: lessons and constraints for water PES in Tanzania. Ecol Econ 69: 1253–1261

Geist HJ, Lambin EF (2004) Dynamic causal patterns of desertification. Bioscience 54:817–829

Gurr GM, Wratten SD, Altieri MA (eds) (2004) Ecological engineering for pest management: advances in habitat manipulation for arthropods. CSIRO, Victoria

Haines-Young R, Potschin M (2010) The links between biodiversity, ecosystem services and human well-being. In: Raffaelli D, Frid C (eds) Ecosystem ecology: a new synthesis. BES ecological reviews series. Cambridge University press, Cambridge

IPBES (2010) Intergovernmental science-policy platform on biodiversity and ecosystem services. http://ipbes.net/

IPCC (2007) Climate Change 2007: synthesis report. Intergovernmental Panel on Climate Change, Geneva

Kumar P (2007) Economics of environment and development. CRC Press

Kumar P (2010) The economics of ecosystems and biodiversity: ecological and economic foundations. TEEB—The Economics of Ecosystems and Biodiversity. Routledge, UK

Kumar M, Kumar P (2008) Valuation of the ecosystem services: a psycho cultural perspective. Ecol Econ 64:808–819

Kumar P, Reddy BS (2007) Ecology and human well-being. Sage Publications Pvt. Ltd, New Delhi

Kumar P, Thiaw I (2013) Values, payments and institutions for ecosystem management: a developing country perspective. United Nations Environment Programme, Nairobi

Maslow AH (1943) A theory of human motivation. Psychol Rev 50:370–396

MEA (2005) Millennium ecosystem assessment synthesis report. Island Press, Washington

Naeem S, Bunker D, Hector A, Loreau M, Perrings C (eds) (2009) Biodiversity, ecosystem functioning, and human wellbeing: an ecological and economic perspective. Oxford University Press, Oxford

Pearce DW, Turner RK (1990) Economics of natural resources and the environment. John Hopkins University Press, Baltimore, USA

Sandhu S (2010) Shifting paradigms in corporate environmentalism: from poachers to gamekeepers. Bus Soc Rev 115:285–310

Sandhu H, Sandhu S (2014) Linking ecosystem services with the constituents of human well-being for poverty alleviation in eastern Himalayas. Ecol Econ 107:65–75

Sandhu HS, Wratten SD, Cullen R, Case B (2008) The future of farming: the value of ecosystem services in conventional and organic arable land. An experimental approach. Ecol Econ 64:835–848

Sandhu H, Wratten S, Costanza R, Pretty J, Porter J, Reganold J (2015) Global significance of non-traded ecosystem services on farmland. PeerJ 3:e762. doi:10.7717/peerj.762

Summer JK, Smith LM, Case JL, Linthrust RA (2012) A review of the elements of human well-being with an emphasis on the contribution of ecosystem services. Ambio 41:327–340

Sen A (2000) Development as Freedom. Anchor Books, New York
Turner KR, Daily GC (2008) The ecosystem services framework and natural capital conservation. Environ Resour Econ 39:25–35
Washington H (2012) Human dependence on nature. Routledge, UK
Wratten S, Sandhu H, Cullen R, Costanza R (eds) (2013) Ecosystem services in agricultural and urban landscapes. Wiley-Blackwell, Oxford

Chapter 10
Sustainable Ecosystems and the Role of Business

Sukhbir Sandhu

10.1 Introduction

There is growing awareness of the intensity and the anthropogenic nature of the environmental change happening on our planet (WCED 1987; MEA 2005; IPCC 2007; UNEP 2010; WBCSD 2010). This is partly due to growing human population coupled with rapid and uninhibited industrial growth to support the demand for various products and services (MEA 2005; UNEP 2010). The environmental degradation has resulted in loss of biodiversity and ecosystem services worldwide (MEA 2005). It is impossible to separate economic development from environmental issues (TEEB 2010). Human activities ranging from primary production such as agriculture to industrial activities that erode the environmental resources upon which they are based (MEA 2005). Environmental degradation in turn can undermine economic development (WCED 1987; WBCSD 2010). Due to growing concern about the environment by governments and people, businesses world over are being pressurised into being more environmentally benign (Jansen and Vellema 2004). Businesses are, therefore, looking for various approaches to integrate ecosystems thinking into business practices.

An increasing number of businesses have recognised the need for achieving a more sustainable form of development through corporate environmental practices (Hart 2007; Sandhu 2010; Sandhu et al. 2010, 2012a). Corporate environmentalism is the recognition of the importance of the natural environment by business organisations and its integration into strategic decision making (Gladwin et al. 1995; Hart 1995; Shrivastava 1995; Banerjee 2002; Dunphy et al. 2007). Businesses are focusing on improving environmentally sustainable practices in response to the increased environmental performance demands from community. This definition includes both compliance based environmental responsiveness and

S. Sandhu (✉)
School of Management, University of South Australia, Adelaide, Australia
e-mail: Sukhbir.Sandhu@unisa.edu.au

H. Sandhu (ed.), *Ecosystem Functions and Management*,
DOI 10.1007/978-3-319-53967-6_10

also extends to voluntary initiatives such as environment management systems, product stewardship and a commitment to sustainable development.

There is a growing body of research that has contributed a lot to our understanding about corporate environmentalism. However corporate environmentalism is a relatively new phenomenon (with most of the research dating back only two decades or so) and is largely restricted to the business practices only. There is lack of a comprehensive framework that can guide businesses to study and implement the principles of corporate environmentalism and move beyond the business realism to integrate socio-economic and environmental benefits into overall sustainable development of the society (IFC 2012).

This chapter builds on the ecosystem services (ES) framework of the Millennium Ecosystem Assessment (MEA 2005) which allows businesses to examine the risks and opportunities that they face with regards to their dependence on natural resources and ecosystem services (TEEB 2010; IFC 2012). This framework is underpinned by the resource dependence theory and incorporates tools to assess their current reliance on ecosystems and their services. The chapter is organised as follows: in the following section, I discuss the concept of ES approach and provide an overview of the application of this approach to business. I then discuss insights from resource dependence theory, which enable business to identify the stakeholders who pose a risk to their operations. I then elaborate how this framework is adapted by a diversified business organisation into their agriculture and forestry sectors. I conclude by discussing the role of ES framework in incorporating sustainability into business management systems.

10.2 Ecosystem Services and Business: Theoretical Framework

Business organisations are responding to the growing demand for sustainable practices by considering a range of environmental management programs (Céspedes-Lorente and Galdeano-Gómez 2004). A new approach to capture, measure, compare and communicate sustainability is by utilising the concept of ES (MEA 2005; Sandhu et al. 2012a). This section builds on the relevance of ES approach to develop a sustainable business.

As discussed in previous chapters, ES are the benefits obtained either directly or indirectly from natural and managed ecological systems (Daily 1997; MEA 2005). These services support life through a wide range of processes and functions and also provide natural capital for large number of industries. Business and ecosystems are intricately linked (Hanson et al. 2008), for example, food industry depends upon land and water resources, pollination, pest control, climate regulation; timber industry depends upon continued supply of timber from forests (Table 10.1). Recent realisation of the loss and declining ecosystems has severe consequences not only for the sustainability but for existence of several industries (Athanas et al. 2006).

Table 10.1 List of ecosystem goods and services associated with natural resource dependent business organisations

	Ecosystem goods and services	Example
1	*Provisioning services*	
	Food	Grains, oil, tea, coffee, vegetables
	Milk	Milk products
	Cocoa	Chocolate products
	Coffee beans	Coffee drink
	Seafood	Shrimps
	Fibre	Cotton
	Seed	Seed crops
	Wood	Timber
	Horticultural products	Fruits
2	*Regulating services*	
	Greenhouse gas regulation	Emission reduction through supply chain management
	Climate regulation	Carbon emission reduction
	Water regulation	Water use efficiency
	Soil erosion control	Payments to upstream users for improving watershed services
3	*Supporting services*	
	Pollination	Horticultural products
	Biological control	Natural enemies, pheromones etc.
	Nutrient cycling	Fertilisers
4	*Cultural services*	
	Eco-tourism	Nature's trail, camping in outback
	Recreation	Visits to national parks, bird watching
	Ethics	Social welfare schemes by business in local communities

These issues are likely to be exacerbated by the complexity of global warming and climate change (IPCC 2007).

Implications of loss of ES and natural capital for the natural resource based industry such as agribusinesses, paper industry, pharmaceuticals are far more direct than for manufacturing industries (Grigg et al. 2009; TEEB 2010). Therefore, there is greater need to understand and implement measures to address these issues in agriculture and related industries (Maloni and Brown 2006). So far, organisations have been responding to the regulatory and market pressures to address the challenges of changing environment. However due to increasing stakeholder pressure, organisations which are more dependent on natural resources now need an explicit focus on incorporating ES into their business decisions (Bleda and Shackley 2008; Hanson et al. 2008; Sandhu et al. 2012a).

Based on the MEA framework, presented in Chap. 1, the framework developed here (Fig. 10.1) maintains that continuous flow of ES depends on the structure or condition of natural capital. The linkages between ES and natural capital are

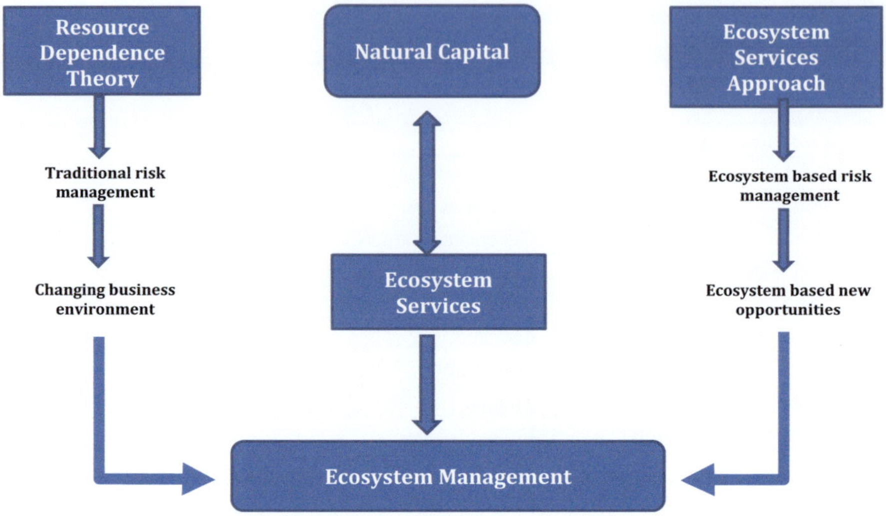

Fig. 10.1 Conceptual framework of business ecosystem management based on the integration of ecosystem services approach and resource dependence theory

impacted by the risks imposed by the natural environment. Natural resource dependence theory helps decision- makers to understand that there is need to apply ES approach in managing ecosystems related risks (Pfeffer and Salancik 1978; Hart 1997; Frooman 1999). According to resource dependence theory, organisations are dependent on stakeholders for their resource needs (Pfeffer and Salancik 1978). It is this dependence on stakeholders for critical resources that gives stakeholders leverage over the firms. One example of this is the recent consumer boycott of Cadbury over the alleged use of slave labour in its contracted agricultural plantation in Africa. The consumer boycott led Cadbury to redefine its mission worldwide to include Fair Trade certified products. Because ecosystems encompasses society and the environment, they are a key stakeholder for land based business organisations and environmental managers need to respond by adopting a carefully considered approach in order to continue their business and maintain the ecosystems, on which they are so intimately reliant.

10.3 Tools for Managing Ecosystem Based Risks

ES are currently in decline (MEA 2005) and are of pressing concern for business that depends on natural resources such as agribusiness and food business. However, business organisations often fail to make connection between ecosystems and their business. But they are also in a unique position to alleviate these challenges through a careful assessment of ES dependence and usage. There are two key tools that can be used to assess risks and business opportunities.

10.3.1 Ecosystem Services Review (ESR)

The World Resources Institute (WRI, Hanson et al. 2008) has developed the corporate Ecosystem Services Review (ESR) tool to link business with ecosystems. It is a strategic tool that helps managers to manage ecosystem related business risks and opportunities. ESR can be done as a stand-alone process or be integrated into the existing environmental management systems of the business. The ESR can provide value to businesses in industries that directly depend on natural resources such as agriculture, beverages, water services, forestry, electricity, oil, gas, mining, and tourism. It is also relevant to sectors such as general retail, healthcare, consulting, financial services, and others to the degree that their suppliers or customers interact directly with ecosystems. General retailers, for example, may face reputational or market risks if some of their suppliers degrade ecosystems and the services they provide. However, ESR is not a quantitative tool and does not help in identifying environmental issues or business opportunities.

10.3.2 Ecosystem Based Business Risks Analysis Tool (EBBRAT)

To overcome some of the issues discussed above, an ecosystem based business risks analysis tool (EBBRAT) has been developed by Sandhu et al. (2012b). The EBBRAT is anchored in the resource dependence theory discussed above. It seeks to qualitatively and quantitatively explore and identify the risks and opportunities that business face because of their dependence on ES and thereby on stakeholders who have a stake in these services and can exploit the firm's dependence on them for critical resources. EBBRAT consists of a MS Excel spread sheet with data entry and output worksheets. A company's dependence and impact on ecosystem services is entered into a data entry sheet. Dependence on a particular ecosystem services relevant to company's performance is measured on a scale, ranging from 0 (which indicates no relevance), to 10 (which is highly relevant). Whereas company's impact is quantified as either positive, negative or no impact on a scale of −10 to +10. A positive impact indicates that the company's operations improve the ecosystem services, whereas a negative impact indicates that it results in decline of ecosystem services. These responses are generated in the output worksheet, and data are automatically organized in the analysis sheet into different dependence categories and different impact categories with different colour schemes. A graph is then generated showing the risks and opportunities based on impact dependence profiles of ecosystem services. This tool also can be used to identify business opportunities where a company can invest and develop risk management options.

10.4 Case Study of an Organisation

Based on the above framework, the following section elaborates a case study of a diversified business organisation that provides sustainable outcomes to its bottom-line, society and the environment. ITC Limited with a market capitalization of over USD 45 billion and revenue of about USD 8 billion is one of India's largest corporations with a presence in tobacco, hotels, paper boards, foods, fashion retailing and commodity export (www.itcportal.com). This organisation is a leading corporate in achieving sustainability in social, economic and environmental dimensions. It employs about 30,000 people and supports the livelihood of about 6 million people indirectly. It is the only business organisation in the world that is water, carbon and solid waste recycling positive (ITC Sustainability Report 2014). Its sustainability program is integrated through all business units. A comprehensive framework includes a multipronged approach to manage natural resources. Sustainable management of natural resources such as water and land results in their protection. Waste recycling and reduction in carbon emissions by increasing efficiencies result in achieving environmental standards. Programs which are spread across different business divisions are aimed at rural or marginal farmers and people living in tribal areas. These programs and projects are aligned with the United Nations Millennium Development Goals (MDGs; UN 2000). See Box 10.1. There are two key areas, where ITC has large interventions programs—one in agriculture sector and one in forestry. Each of these are discussed below.

Box 10.1 Corporate Social Responsibility at ITC

Millennium Development Goals	Activities at ITC
Eradicating extreme poverty and hunger	Agricultural development, animal husbandry, Agricultural extension activities
Universal primary education	Promoting and supporting primary education
Promote gender equality	Promoting women's micro enterprises Sustainable livelihoods for women
Global partnership for development	Promotions of art culture, partnering with global organisations for skill development
Ensure environmental sustainability	Social forestry, rural development, natural resources conservation and management

10.4.1 Sustainable Ecosystems in Agriculture Sector

ITC's International Business Division (IBD) is engaged in exports of a range of agricultural commodities. ITC initiated efforts to re-engineer procurement of farm products by introducing information and technology into rural villages through a project widely known as *e-chaupal* (Annamalai and Rao 2003). About 40,000 villages have *e-chaupal* kiosks and about 4 million farmers benefit from it.

It has brought many advantages which extend beyond good remuneration to the farmer. This network has helped rural transformation by providing more transparency about the market to the farmers. It has helped to improve farmers' productivity and incomes in about 100,000 villages across India.

It requires identification and training of a local farmer who acts as *Sanchalak*, to manage each e-Choupal. *Sanchalak*'s hosts a computer at his house which is linked to internet. This whole setup costs between USD 3000 and 6000. Farmers are not charged for using the system. However, *Sanchalak* who serve the community benefits by increased prestige and a small commission paid for each transaction. Farmers access daily rates of the commodities and can decide when to sell their produce through this system. ITC can buy the crop directly from the farmer, bring it to the ITC complex known as *Chaupal Sagar* (Fig. 10.2), usually at the District level, weigh it and check for quality. The farmer is then paid for the value of crop.

Farmers can also order all farm inputs and household products through this system as the prices are generally lower than the village trader due to the partnerships of ITC with consumer goods companies.

This way farmers are protected from the traditional traders who indulge in several malpractices leading to economic losses to farmers. This system is transforming agriculture sector by improving social and economic conditions of the rural households especially farmers. This project has also resulted in significant ecosystem improvements, due to much more careful and informed approach to water usage, pesticide applications, and reduced crop wastage.

Fig. 10.2 Photos showing ITC Ltd. *Chaupal Sagar* providing agricultural information and inputs alongwith consumer goods in *e-chaupal* project. *Photos by* S. Sandhu

10.4.2 Social Forestry

ITC Plantation and Specialty Papers Division is also engaged in the development of rural communities through the extension of farm forestry and social forestry models. Small and marginal farmers living in remote areas are being involved in this project. There are about 200,000 ha under social and farm forestry model in 2400 villages. Farmers provide their land and the company provides all the technical knowhow and continued extension support to grow eucalyptus trees on the farmland. There are two models prepared by ITC; farm forestry (Fig. 10.3) and agroforestry (Fig. 10.4).

Fig. 10.3 Farmer showing his agroforestry plot with eucalyptus trees for paper mill. In between row area is used for other crops cultivation (For example, cotton, pulses, etc.). *Photos by* S. Sandhu

Fig. 10.4 Agroforestry model showing eucalyptus trees at a farmer's field. *Photos by* S. Sandhu

In farm forestry model, farmers plant eucalyptus saplings provide by the ITC extension division at 1.5 × 3 m distance with 2200 trees per hectare. These are grown in areas, which receive very less rainfall, and generally crop cultivation is not possible. Trees mature in about four years yielding on an average 100 tons of timber per hectare. There is generally a buyback program under which ITC mills buy timber from the farmers at market rates. On an average farmers receive USD 74 per ton of timber resulting in gross income of USD 7370 per hectare. Tree harvesting and bark removal and transport to the mill costs about USD 15 per ton of timber. This leaves a net income of about Rs. 5896 per hectare after four years to farmers (giving farmers a net income of USD 1475 per hectare per year). This income source has drastically improved the socio-economic status of marginal farmers in poor rural areas. There is improvement of associated social indicators such as children's education, women's health, better houses and generally a higher standard of living in the region. This project is currently going on in three districts in Andhra Pradesh.

In agroforestry models, there is different spacing of trees with 8 m wide row between two rows of eucalyptus trees. This leaves sufficient space to grow other crops provided there are irrigation facilities available. Farmers grow cotton, chilli, tobacco and pulses, etc. This model is not only supplying timber and providing lump sum income to farmers after four year cycle but also provide opportunities to generate cash income through the year.

ITC paper division is also setting up biodiversity plots in this region to protect local flora and faunal diversity (Fig. 10.5). There are total nine biodiversity plots with a total area of 56 hectares in Andhra Pradesh. Thus, agro forestry and farm

Fig. 10.5 Biodiversity plots within farm area to conserve local flora and fauna under the *Mission Sunehra Kal* programme. *Photos by* S. Sandhu

forestry scheme is bringing rural transformation in this region. While there are clear benefits to the organisation (e.g., a continuous supply of timber), there are also significant environmental benefits of protecting virgin forests through growing timber in wastelands. As discussed above, there are also social benefits that accrue to the farmers. Thus, a carefully developed ecosystem based approach can result in economic, social and environmental sustainability.

The programs described above are part of the social investment mission known as *Mission Sunehra Kal* at ITC (Fig. 10.5). It involves managing social and natural capital to create value to all sections of the society (ITC Sustainability Report 2014). The aim of this mission is to build on the ecosystem based approach discussed in the book to empower all stakeholders to promote sustainable livelihoods.

10.5 Summary

Increasingly, business organisations that are dependent on natural resources are adopting more eco-friendly practices. Here, a holistic framework is presented that provides tools and methods to asses risks and develop opportunities for the business organisations. It has potential to encourage organisations to create tangible business outcomes, while reducing ecosystem impact. As business organisations become constrained by natural resources and climate risks, there is high potential for the adoption of this approach to reduce risks and find sustained competitive advantage.

References

Annamalai K, Rao S (2003) What works: ITC's e-choupal and profitable rural transformation. World Resources Institute, Washington DC. http://www.wri.org/publication/what-works-7

Athanas A, Bishop J, Cassara A, Donaubauer P, Perceval C, Rafiq M, Ranganathan J, Risgaard P (2006) Ecosystem challenges and business implications. Earthwatch Institute, World Resources Institute, WBCSD and World Conservation Union

Banerjee SB (2002) Corporate environmentalism: the construct and its measurement. J Bus Res 55:177–191

Bleda M, Shackley S (2008) The dynamics of belief in climate change and its risks in business organizations. Ecol Econ 66:517–532

Céspedes-Lorente J, Galdeano-Gómez E (2004) Environmental practices and the value added of horticultural firms. Bus Strategy Environ 13:403–414

Daily GC (ed) (1997) Nature's Services: Societal Dependence on Natural Ecosystems. Island Press, Washington, D.C.

Dunphy D, Griffiths A, Benn S (2007) Organizational change for corporate sustainability, 2nd edn. Routledge, London

Frooman J (1999) Stakeholder influence strategies. Acad Manag Rev 24:191–205

Gladwin TN, Kennelly JJ, Krause TS (1995) Shifting paradigms for sustainable development: implications for management theory and research. Acad Manag Rev 20:874–908

Grigg A, Cullen Z, Foxall J, Strumpf R (2009) Linking shareholder and natural value. Managing biodiversity and ecosystem services risk in companies with an agricultural supply chain. Fauna

& Flora International, United Nations Environment Programme Finance Initiative and Fundação Getulio Vargas. http://www.naturalvalueinitiative.org/download/documents/Publications/LSNVExecSummary.pdf

Hanson C, Ranganathan J, Iceland C, Finisdore J (2008) The corporate ecosystem services review: guidelines for identifying business risks and opportunities arising from ecosystem change. World Resources Institute, World Business Council for Sustainable Development, Meridian Institute, pp 48

Hart SL (1995) A natural resource based view of the firm. Acad Manag Rev 20:986–1014

Hart S (1997) Beyond greening: strategies for a sustainable world. Harv Bus Rev, January–February, 66–76

Hart SL (2007) Capitalism at the crossroads: aligning business, Earth and Humanity. Wharton School Publishing, New Jersey

IFC (2012) Policy and performance standards on environmental and social sustainability. International Finance Corporation, World Bank, Washington, D.C.

IPCC (2007) Climate change 2007: The physical science basis. http://ipcc-wg1.ucar.edu/wg1/docs/WG1AR4_SPM_Approved_05Feb.pdf

ITC Sustainability report (2014) For all our tomorrows. http://www.itcportal.mobi/sustainability/sustainability-report-2014/index.aspx#

Jansen K, Vellema S (2004) Agribusiness and society: corporate responses to environmentalism, market opportunities and public regulation. Zed Books, London

Maloni MJ, Brown ME (2006) Corporate social responsibility in the supply chain: An application in the food industry. J Bus Ethics 68:35–52

MEA (2005) Ecosystem and human well-being: synthesis. Millennium ecosystem assessment. Island Press, Washington, D.C.

Pfeffer J, Salancik GR (1978) The external control of organizations. Harper and Row, New York

Sandhu S (2010) Shifting paradigms in corporate environmentalism: From poachers to gamekeepers. Bus Soc Rev 115:285–310

Sandhu SK, Ozanne L, Smallman C, Cullen R (2010) Consumer driven corporate environmentalism: Fact or fiction? Bus Strategy Environ 19:356–366

Sandhu S, Smallman C, Ozanne LK, Cullen R (2012a) Corporate environmental responsiveness in India: Lessons from a developing country. J Clean Prod 35:203–213

Sandhu H, Nidumolu U, Sandhu S (2012b) Assessing risks and opportunities arising from ecosystem change in primary industries using ecosystem-based business risk analysis tool. Hum Ecol Risk Assess 18:47–68

Shrivastava P (1995) The role of corporations in achieving ecological sustainability. Acad Manag Rev 20:936–960

TEEB (2010) The economics of ecosystems and biodiversity report for business—Executive summary. http://www.teebweb.org

UN (2000) Millennium development goals. http://www.un.org/millenniumgoals

UNEP (2010) The role of ecosystems in developing a sustainable 'green economy' UNEP Policy Series, Brief 2

WBCSD (2010) Vision 2050: the new agenda for business. The World Business Council for Sustainable Development, Switzerland

WCED (1987) Our common future. Oxford University Press, New York

Epilogue

Designing Sustainable Ecosystems for Equitable and Sustainable Development

Harpinder Sandhu
School of the Environment, Flinders University, Adelaide, Australia.

It is globally accepted that society and economic systems both depend on healthy ecosystems for various benefits widely known as ecosystem services (ES). These life supporting ES are in a state of decline worldwide due to human activities that include mismanagement of ecosystems. There is sufficient scientific evidence to suggest policy and practice to include ecosystem-based management to achieve three-dimensional sustainability, i.e., social, ecological and economic. However, governments, private sector and community at large are unable to respond in an integrated manner to address these multifaceted challenges due to loss of biodiversity and ecosystems. This is despite the acknowledgement that livelihoods and well-being depends on the state of ecosystems that supply various industrial raw material and other cultural and aesthetics benefits to human beings. Global action led by the United Nations to understand the role of ecosystems and their services in social and economic systems have raised awareness of the critical nature of the environmental challenges. However, there is need to understand and fulfill the gap between scientific evidence and the policy for the long-term sustainability. This book highlights the role played by ecosystems in South Asian region. Human population is growing at an increasing rate in the region, therefore, the impacts on natural resources on ecosystems will be enormous in coming years. To tackle these issues, an ecosystem-based approach, discussed in the book, will be required to resolve some of the socio-economic and environmental issues in the region.

Ecosystems have generally porous boundaries and several underlying functions are required for them to produce certain benefits. For example, agroecosystems are dependent on freshwater systems for water; urban ecosystems are dependent on all surrounding systems for food, clean air, water supply etc. Similarly, with the realisation that these ecosystems are of high economic value, it is difficult to oversee their role in providing human welfare, directly or indirectly. It is not a good approach to seek solutions independent of each other. A comprehensive and integrated approach as discussed in the book is required for desirable outcomes. A pre-requisite is an integrated effort by all segments so the society to look holistically. This must be weaved into the local, national and regional policies in

H. Sandhu (ed.), *Ecosystem Functions and Management*,
DOI 10.1007/978-3-319-53967-6

South Asia. There is need to account this social and natural capital and ES in the national accounts so that a thorough balance sheet can be prepared and maintained every year. This also requires a collective vision to develop equitable and sustainable society.

There is greater need to incorporate the essence of human dependence on nature and ecosystems in education sector at primary, secondary and tertiary level and for all disciplines. A well informed and better-educated society can think beyond the current realm of silos approach to seek transdisciplinary solutions and maintain balance with the natural environment. Local, national and regional community then can institutionalise these into their day-to-day life. A shift in policy at macro and micro level is required to bring this change by understanding the role and value of ecosystems in providing human well-being. Such policy changes can shift the management thinking from problem to solutions oriented approaches to yield sustainable and desirable future for all segments of the society.

South Asia has rich history and culture dating back to thousands of years. Many great civilisations flourished in the region and gave rise to an educated and benign society that had great appreciation for the nature and its resources. Industrialisation in recent decades along with increasing population and its demand for food and consumer products have resulted in faster rate of environmental degradation. This rate is likely to be complicated and exacerbated by the climate shift, which is already showing its impact, unless an alternative sustainable development approach is adopted by the society. Discounting nature over economic growth is certainly not an option for the region as millions depend on these ecosystems and their services directly and indirectly. We know that not only society but economy also depend on these ecosystems, which are of high economic value. It is worth taking a pause to reflect, which direction are we heading—in pursuit of, an endless goal of amassing material wealth or a healthy and fulfilled life. What is your goal in life?

December 2016 Harpinder Sandhu (Adelaide)

MIX
Papier aus verantwortungsvollen Quellen
Paper from responsible sources
FSC® C105338

If you have any concerns about our products,
you can contact us on
ProductSafety@springernature.com

In case Publisher is established outside the EU,
the EU authorized representative is:
Springer Nature Customer Service Center GmbH
Europaplatz 3, 69115 Heidelberg, Germany

Printed by Libri Plureos GmbH
in Hamburg, Germany